driving
buses and coaches
the official DSA syllabus

Approved by Plain English Campaign

London: TSO

Published with the permission of the Driving Standards Agency on behalf of the Controller of Her Majesty's Stationery Office.

© Crown Copyright 2003

Applications for reproduction should be made in writing to
Commercial Department, Driving Standards Agency, Stanley House, 56 Talbot Street, Nottingham NG1 5GU

First edition Crown copyright 1995
Second edition Crown copyright 1997
Third edition Crown copyright 1999
Fourth edition Crown copyright 2001
Fifth edition Crown copyright 2002
Sixth edition Crown copyright 2003
Third impression 2004

ISBN 0 11 552486 X

A CIP catalogue record for this book is available from the British Library

Other titles in the official Driving Series

The Official Theory Test for Drivers of Large Vehicles

The Official Theory Test for Drivers of Large Vehicles (CD-ROM)

Driving Goods Vehicles – the Official DSA Syllabus

The Official Guide to Tractor and Specialist Vehicle Driving Tests

The Official Theory Test for Car Drivers

The Official Theory Test for Car Drivers and Motorcyclists (CD-ROM)

The Official Theory Test Practice Papers for Car Drivers

The Official Driving Test

Driving – the Essential Skills

The Official Theory Test for Motorcyclists

The Official Theory Test Practice Papers for Motorcyclists

Official Motorcycling – CBT, Theory and Practical Test

Motorcycle Riding – the Essential Skills

The Official Guide to Accompanying Learner Drivers

Printed in the United Kingdom by TSO
N172166   08/04   C20   989579   19585

ABOUT THE DRIVING STANDARDS AGENCY

The Driving Standards Agency (DSA) is an executive agency of the Department for Transport (D*f*T).

You'll see its logo at test centres.

DSA
DRIVING STANDARDS AGENCY
SAFE DRIVING FOR LIFE

DSA aims to promote road safety through the advancement of driving standards, by

- establishing and developing high standards and best practice in driving and riding on the road; before people start to drive, as they learn, and after they pass their test
- ensuring high standards of instruction for different types of driver and rider
- conducting the statutory theory and practical tests efficiently, fairly and consistently across the country
- providing a centre of excellence for driver training and driving standards
- developing a range of publications and other publicity material designed to promote safe driving for life.

DSA website

www.driving-tests.co.uk

DfT website

www.dft.gov.uk

Green issues website

www.defra.gov.uk/environment/index.htm

Acknowledgments

The Driving Standards Agency (DSA) would like to thank the following organisations for their contribution to the production of this publication

Buckinghamshire Fire and Rescue Service

The Confederation of Passenger Transport UK

County Durham and Darlington Fire and Rescue Brigade

Department for Transport

ETA Services Ltd

The National Federation of Bus Users

The Traffic Director for London

Transport Licensing and Enforcement Branch, Department of Environment Northern Ireland

The Vehicle Inspectorate Test Station in Norwich and Dereham Coachways (Norfolk) for their kind cooperation in securing the front cover photograph

Every effort has been made to ensure that the information contained in this publication is accurate at the time of going to press. The Stationery Office cannot be held responsible for any inaccuracies. All metric and imperial conversions in this book are approximate. Please check that you have the most up-to-date version of this publication that is available.

Information and advice contained within this publication is for guidance only. Vehicle maintenance or other tasks must be carried out only with careful reference to an individual vehicle's handbook and any safety information relating to the vehicle. Close regard must also be had to the health and safety of those undertaking such tasks and any others who might be affected by undertaking such tasks.

Please note that individual commercial organisations may prohibit their employees from carrying out some or all of the maintenance or other tasks described in this publication and users are advised to check with their employers before attempting such tasks.

CONTENTS

About the Driving Standards Agency   iii
Acknowledgments   iv
Introduction   1
About this book   2

PART ONE   Getting started

- Applying for your licence   5
- The theory test   6
- Eyesight requirements   7
- Medical examination and form D4   8
- Professional standards   10
- Responsibility   11
- Attitude   12
- Passengers   17

PART TWO   Passenger carrying vehicles

- Forces at work   25
- Maintaining control   29
- Vehicle sympathy   30
- Types of PCV   31
- Vehicle maintenance   46
- Other auxiliary systems   55

PART THREE   Limits and regulations

- Basic knowledge   57
- Environmental issues   64
- Drivers' hours   68
- Other regulations   73
- Anti-theft measures   87

PART FOUR   Driving skills

- Professional driving   89
- Driving at night   100
- Motorway driving   106
- All-weather driving   122
- Breakdowns   129
- Accidents   132
- First aid   138

PART FIVE   Test preparation

- Preparing for the driving test   143
- Applying for the test   151

v

- The official syllabus — 156
- Revised legislation — 169

PART SIX   The PCV driving test

- What to expect on the day — 171
- Safety checks — 175
- The reversing exercise — 177
- The braking exercise — 180
- The vehicle controls — 182
- The gear-changing exercise — 191
- Other controls — 192
- Moving off — 193
- Using the mirrors — 194
- Giving signals — 198
- Acting on signs and signals — 199
- Making progress — 200
- Controlling your speed — 201
- Separation distance — 202
- Awareness and anticipation — 203
- Hazards — 203
- Selecting a safe place to stop — 222
- Uncoupling and recoupling — 223
- Understanding the rules — 225
- The test results — 226

PART SEVEN   Additional information

- Disqualified drivers — 229
- DSA services — 231
- Useful addresses — 233
- PCV licence entitlements — 239
- Minimum test vehicles — 240
- New MTV requirements — 241
- Vehicle types and licence requirements — 242
- Cone positions — 243
- Road signs — 244
- Conclusion — 245
- Glossary — 246

INDEX — 250

INTRODUCTION

This book provides detailed professional guidance for the safe driving of buses and coaches. As the driver of a vehicle carrying passengers you must accept responsibility for their safety. Whether you drive a minibus with eight passengers or a double-deck coach with 88, each one is relying on you to look after them.

The starting point for a professional driver is having the correct attitude. You should set an example through courtesy and consideration, and make allowances for the mistakes of other drivers. A professional driver will have a sound knowledge of driving theory, coupled with the ability to apply that theory in an expert manner.

Changes to the driving test have resulted from new legislation required by the European Commission (second EC directive), and implemented in the interest of road safety. These changes are included in this book to help PCV drivers to keep up to date with the new legislation.

This book provides an officially recommended syllabus for learning how to drive a bus or coach, plus a structured approach to training that should help you to progress to a professional standard. Studying *Driving Buses and Coaches – the Official DSA Syllabus* will help you to achieve a better understanding of the skills and attitudes that combine to make for higher driving standards.

Robin Cummins
**The Chief Driving Examiner
Driving Standards Agency**

ABOUT THIS BOOK

This book will help you to

- understand what is expected of a passenger carrying vehicle (PCV) driver
- prepare for your practical PCV driving test.

Part One tells you how to obtain a provisional PCV licence, together with the essential differences between driving cars and larger vehicles. The importance of the correct attitude towards driving and passengers is also covered.

Part Two includes information about different types of PCVs. Their driving characteristics are explained.

Part Three explains what PCV drivers need to know about the legal limits that affect their vehicle and their driving. It also deals with the environmental issues and what you can do to help.

Part Four gives advice on professional driving, as well as driving in various weather conditions, at night and on motorways. It also deals with accidents and breakdowns and includes useful advice on first aid.

Part Five contains the officially recommended syllabus for the PCV driving test. There's also detailed guidance on how to prepare and apply for the test.

Part Six explains all the aspects of the driving test. The test requirements, the skills you should show and the faults you should avoid are explained fully. Action you can take upon receiving your test results is also considered.

Part Seven outlines DSA services, including the complaints guide and compensation code. Useful addresses and other helpful information can also be found here.

The important factors

Reading this book should help you to appreciate the principles of driving PCVs and so lead you to become a safer driver. However, this book is only one of the important factors in your training. The others are

- **good instructor**
- **plenty of practice**
- **your attitude.**

Once you have obtained your PCV licence you should take pride in your driving. Your professionalism will be seen and appreciated by other road users.

Driving is a life skill

ABOUT THIS BOOK

Study Materials

You should already have a good, sound knowledge of driving skills. It's strongly recommended that you study a current copy of *The Highway Code*. You can buy one from any good bookshop or newsagent. *The Highway Code* is now also available with the *Theory Test Companion*, which summarises the contents of the theory test for car drivers. It can also be found on a dedicated website, www.highwaycode.gov.uk.

The DSA series of books (see page ii) will provide you with the information you need to further your skills and knowledge. *Driving - the Essential Skills* and *The Official Theory Test for Drivers of Large Vehicles* are particularly recommended.

A DVD entitled *The Official guide to Hazard Perception* for all drivers and riders will help candidates prepare for the hazard perception part of the theory test. A video and workbook-based training pack entitled *Roadsense* is also available. These are very useful tools for existing drivers wanting to brush up on their hazard awareness skills.

All the above publications can be bought from some retail outlets and from DSA Merchandising, tel. 0870 241 4523.

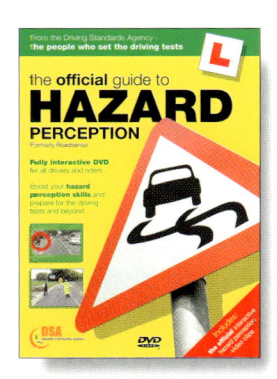

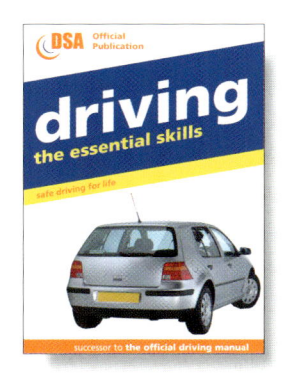

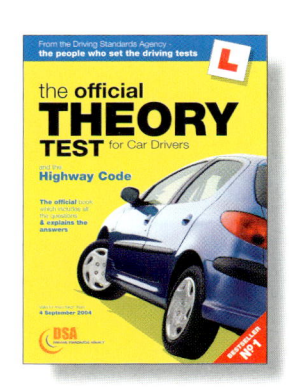

Part One

Getting started

The topics covered

- **Applying for your licence**
- **The theory test**
- **Eyesight requirements**
- **Medical examination and form D4**
- **Professional standards**
- **Responsibility**
- **Attitude**
- **Passengers**

GETTING STARTED

Applying for your licence

In order to drive PCVs you must apply to the Driver and Vehicle Licensing Agency (DVLA), Swansea, for provisional PCV entitlement to be added to your full car (category B) licence, unless you've held the entitlement before. You must not drive a PCV until you've received your licence with the proper category added.

To be issued a licence to drive PCVs you must

- be at or above the minimum age for the category of vehicle you intend to drive – see the relevant table in Part Seven
- have a full category B licence – you can't take a driving test on a bus before passing a car test
- meet the medical requirements
- pay the fee.

You can obtain a driving licence application form (D1) from

- Post offices
- Traffic Area Offices
- Vehicle registration offices (VROs)
- DVLA Customer Enquiries Unit, Swansea, SA6 7JL.

Read carefully the notes that accompany the form and fill in all the relevant parts. The form may have to be returned to you, and there could be a delay in issuing your licence if you leave anything out. If you need advice about completing the form, ring the enquiry number for DVLA listed at the back of this book.

Send the completed application form and

- your category B licence, or provisional car driving licence and a valid driving test pass certificate (form D10)
- a completed medical report form (D4), signed by a doctor
- the appropriate fee

to DVLA, Swansea, SA99 1BR. The current fees and ways to pay are listed on the application form.

Automatic transmission

If your category B licence is restricted to automatic vehicles you can only drive a PCV with an automatic gearbox. Vehicles with fully automatic, semi-automatic, pneumo-cyclic, electronically controlled or pre-select gearboxes with a gear-change pedal are all classified as automatic.

If a vehicle has a clutch pedal it's classified as having a manual gearbox. To drive a PCV with a manual gearbox you'll have to pass a test in a bus or coach with a manual gearbox.

GETTING STARTED

The theory test

All new drivers wishing to drive PCVs will first have to pass a PCV theory test before taking the practical driving test. You can commence your lessons before passing the theory test, but you must pass before a booking for a practical test can be accepted. The theory test pass certificate has a two-year life. If the practical test isn't passed within that time the theory test will have to be retaken and passed before a booking for a practical test can be accepted.

Drivers who already hold a full car licence (category B) won't have to take the theory test to drive minibuses with fewer than nine passenger seats. However, to obtain a category D1 licence to drive a minibus with more than 8 and not more than 16 passenger seats a driver must have a full category B licence and pass the PCV theory test before applying to take the PCV practical driving test. Similarly, to obtain a category D licence to drive a minibus, midibus, coach or single- or double-decker bus with more than 16 passenger seats a driver must have a full category B licence and pass the PCV theory test before applying to take the PCV practical driving test.

Additional notes

- A category D licence is required to drive articulated buses (described in the UK as 'bendi-buses').
- It won't be necessary to obtain a category D1 or D1 + E licence before applying for a category D test.
- If a trailer with a maximum permitted weight of more than 750 kg is to be towed, a category B + E, D1 + E or D + E licence may be required, as appropriate. A trailer weighing less than 750 kg may be towed by any of the vehicles in these licence categories.
- Passengers may NOT be carried in any trailer.

If you're unsure of the category of licence entitlement that you need, see the charts in Part Seven at the back of this book.

As a learner you must be accompanied by a qualified driver who has held a full licence for the category of vehicle being driven for at least three years.

GETTING STARTED

Eyesight requirements

All drivers, for whatever category of vehicle, must be able to read a number plate in good daylight at 20.5 metres (67 feet), or 20 metres (about 66 feet) if the new narrow font letters have been used on the number plate. If glasses or contact lenses are needed to do this, then they must be worn when driving.

In addition, any applicant for a PCV licence must have a visual acuity of at least

- 6/9 in the better eye
- 6/12 in the other eye

when wearing glasses or contact lenses, if needed. There must also be normal vision in both eyes (defined as a 120° field) and no evidence of double vision (diplopia). Satisfactory uncorrected visual acuity is also required of applicants.

All applicants must have an uncorrected visual acuity of at least 3/60 in each eye. This visual field requirement is the normal binocular field of vision.

Your doctor will use the standard Snellen test card to test your eyesight. If you only have eyesight in one eye you must declare this on form D1.

A licence-holder who held a PCV or LGV licence before 1 January 1997 and whose eyesight doesn't meet the required new standard should contact

Drivers' Medical Group
DVLA
Swansea
SA99 1TU

Tel: 0870 600 0301

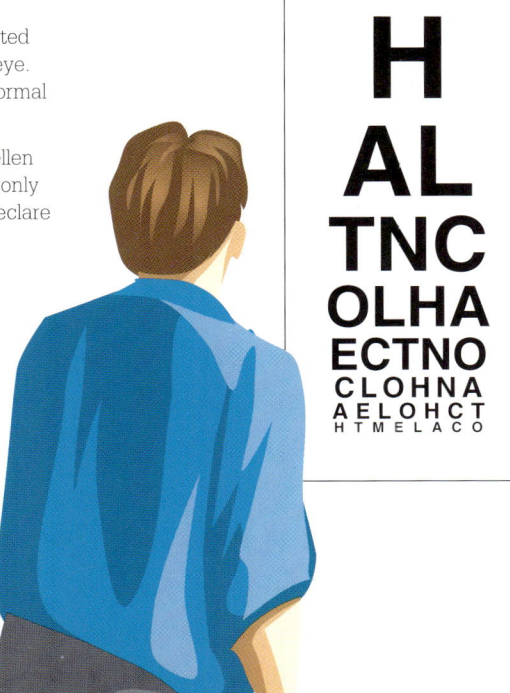

GETTING STARTED

Medical examination and form D4

Consult your doctor first if you have any doubts about your fitness. In any case, if this is your first application for PCV entitlement, a medical examination must be carried out by a doctor.

You'll need to send a medical report (form D4) off with your application. You'll also need to send a report if you're renewing your PCV licence and you're aged 45 or over, unless you've already sent one to DVLA during the last 12 months.

In order to complete form D4 you'll need to undergo a medical examination. You should only complete the applicant details and declaration (Section 8 on the form) when you're with your doctor at the time of the examination. Your doctor has to witness you doing this. The other sections on the form will be completed by your doctor. The medical report will cover

- vision
- nervous system
- diabetes mellitus
- psychiatric illness
- general health
- cardiac health
- medical practitioner details.

After the medical examination, study the notes on pages 1 and 2 of the form. Remove these pages before sending in your application and keep them for future reference.

The medical examination isn't available free under National Health rules. Your doctor is entitled to charge the current fee for this medical examination, which you'll be responsible for. It can't be recovered from DVLA, and the fee isn't refundable if your application is refused.

The completed form must be received by DVLA within four months of your doctor signing it.

GETTING STARTED

Medical standards

You may be refused a PCV driving licence if you suffer from any of the following

- liability to epilepsy/seizure*
- diabetes requiring insulin (unless you held a licence on 1 April 1991 and the Traffic Commissioner who issued that licence had knowledge of your condition)
- eyesight defects (see eyesight requirements on page 7)
- heart disorders
- persistent high blood pressure (see notes on form D4)
- a stroke within the past year
- unconscious lapses within the last five years
- any disorder causing vertigo within the last year
- severe head injury, with serious continuing after-effects, or major brain surgery
- Parkinson's disease, multiple sclerosis or other chronic nervous disorders likely to affect the use of the limbs
- mental disorders
- alcohol/drug problems
- serious difficulty in communicating by telephone in an emergency.

An applicant or licence-holder failing to meet the epilepsy, diabetes or eyesight regulations must by law be refused a licence.

Note
A driver who remains seizure-free for at least 10 years (without anticonvulsant treatment within that time) may be eligible for a licence, but with restricted entitlement. Contact DVLA for further information.

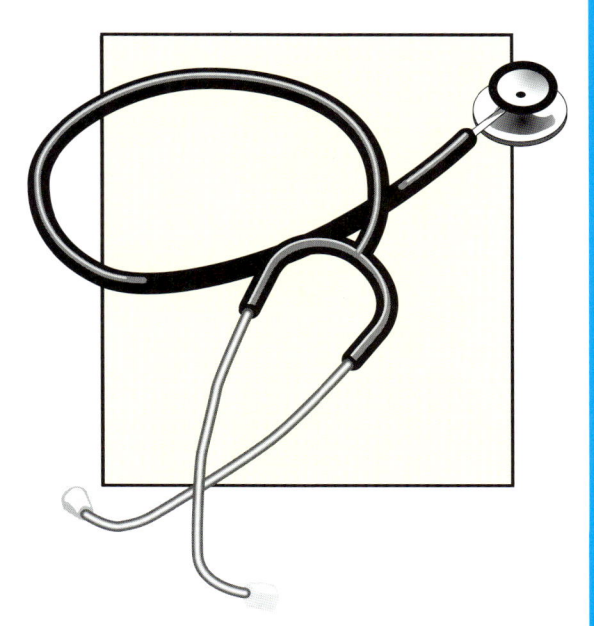

GETTING STARTED

Professional standards

In order to drive a bus, coach or minibus safely you'll need

- a comprehensive knowledge of *The Highway Code*, including the meaning of traffic signs and road markings (especially those that indicate restrictions for large vehicles)
- a thorough knowledge of the regulations that apply to your work
- a high level of driving skill
- the ability to plan well ahead.

You must appreciate the differences between driving larger and smaller vehicles. Some of these aspects will be obvious from the moment you first start to drive a larger vehicle. Other features will only become apparent as you gain more experience.

Always apply the professional driving techniques described in this book. Remember, you must never allow safety to be put at risk.

No risk is ever justified.

GETTING STARTED

Responsibility

As well as your passengers, you must show responsibility towards all other road users. If you act hastily you risk endangering others.

- Drive properly, and your passengers will arrive safely at their destination.
- Drive carelessly or dangerously, and you risk the safety of your passengers and other road users.

When a bus or coach is involved in an accident it's bound to lead to damage, injury or loss of life. As a professional driver you have a part to play in making sure accidents don't happen.

Human error is the main cause of most collisions on the road. High-quality training should help you to avoid making such errors and reduce the risk of you being involved in accidents.

Sometimes accidents are due to the mechanical failure of vehicle components. The way you drive can affect the life of these components. Drivers who demonstrate a high degree of expertise reduce the risk of accidents happening. So, be responsible for driving your vehicle safely and sensibly at all times.

11

GETTING STARTED

Attitude

The sheer size, noise and appearance of a PCV can be intimidating to cyclists, pedestrians and even car drivers. Never use the size of your vehicle in an aggressive way. The general public tend to see the bus or coach driver either as

- a skilful professional who manoeuvres a large vehicle in difficult spaces and delivers passengers safely to their destination, or
- an impatient person determined to make other road users, and his or her passengers, do precisely what he or she wants.

A PCV driver should create the best possible image by setting a good example for others to follow.

Driving large vehicles can be very enjoyable – even more so when you can be proud that you are doing it well.

GETTING STARTED

Tailgating

Tailgating means following dangerously close behind another vehicle, at speed, maybe only a few feet apart. It often happens on motorways.

Not only are tailgating and driving in close convoy with other PCVs bad driving habits, but they often have serious consequences. Some police forces are so concerned at the number of accidents involving tailgating vehicles that they now video the offence and prosecute offenders.

If you are tailgating, your view of the road ahead is seriously restricted and you have an impossible stopping distance.

If a vehicle in front brakes heavily you need time to react and move your foot to the brake pedal. At 50 mph you'll have travelled 15 metres (49 feet – more than the length of a coach) before you start to brake. During that time the vehicle in front could have reduced its speed to below 40 mph.

Always maintain your safety margins. Considerate drivers also allow the drivers following them ample time to react.

GETTING STARTED

Intimidation

Never use the size, weight and power of your vehicle to intimidate other road users. Even the repeated 'hiss' of air brakes being applied or released while stationary gives the impression of 'breathing down the neck' of the driver in front.

Speed

You can never justify driving too fast just because you have to reach a given location by a specific time. Don't be tempted to drive faster when you've fallen behind schedule. If an accident happens and you injure someone there's no possible defence for your actions.

Road speed limiters

Vehicles which require speed limiters

- A coach or bus which was first used on or after 1 April 1974 and before 1 January 1988 and would, if a speed limiter were not fitted, have a relevant speed exceeding 112.65 kph (70 mph).
- A coach or bus which was first used on or after 1 January 1988, has a gross weight exceeding 7500 kg and would, if a speed limiter were not fitted, have a relevant speed exceeding 100 kph (62 mph).

The speed at which the limiter is set must be shown on a plate displayed in a conspicuous position in the driver's cab.

Types

There are two main types of speed limiters. One type works by the mechanical or electrical actuator, the other works through the vehicle's engine management system.

Principles of operation

The speed limiter works by receiving a road speed signal either from the tachograph or a sensor fitted to another system on the vehicle, such as the anti-lock braking system (ABS). Occasionally a specific sensor for the speed limiter system may be fitted. The vast majority of vehicles are fitted with speed limiters that take the speed signal from the tachograph.

Irrespective of the type of sensor used, the information is transmitted to the electronic control unit (ECU), which, in turn, controls the equipment used to regulate the power output or revolutions of the vehicle's engine. This is normally achieved by reducing the amount of fuel supplied to the engine.

Parts

The system will consist of a road speed sensor (this may or may not be part of the tachograph system), an electronic cable, an

GETTING STARTED

electronic control unit (this may or may not be part of the vehicle's engine management system), an actuation device (this may be a pump, relay or valve) and a plate that is fitted to the vehicle to show the set speed.

Connections

Authorised speed limiter centres can only carry out installation, repairs and calibration. These centres will seal all connections between the speed sensor, electronic control unit and the actuation device to ensure the system is tamper-proof.

Maintenance

No day-to-day maintenance is required, although any failure of the road speed limiter must be reported to the operator of the vehicle, who should arrange for the repair at the end of the journey the vehicle is engaged upon.

As a result of the fitting of speed limiters set to 100 kph (62 mph), buses and coaches may not use the far right-hand lane on a motorway. The restriction does not apply to A class roads with three lanes.

Retaliation

You must resist the temptation to retaliate in order to 'teach someone a lesson'. Always drive

- courteously
- with anticipation
- calmly, allowing for other road users' mistakes
- in full control of your vehicle.

You can't act hastily when driving a PCV without the possibility of serious loss of vehicle control.

The horn

PCVs are often equipped with powerful horns and their use should be strictly confined to the guidance set out in *The Highway Code* – to warn other road users of your presence.

Never use the horn

- aggressively
- between 11.30 pm and 7 am in a built-up area
- when stationary, unless a moving vehicle poses a danger.

The headlights

There's only one official use of flashing the headlights described in *The Highway Code*: to let other road users know that you're there.

- Never repeatedly flash the headlights while driving directly behind another vehicle.
- To avoid dazzle, don't put headlights on to full beam when following another vehicle.
- Don't switch on auxiliary lights fitted to your vehicle unless weather conditions require them, and remember to switch them off when the conditions improve.

15

GETTING STARTED

Misleading signals

Neither the horn nor the headlights should be used to rebuke or to intimidate another road user. By using unauthorised 'codes' of headlight or indicator flashing you may be misunderstood by others. This in turn could lead to an accident.

When driving abroad, headlamp flashing is used purely as a warning. Any other intention won't be understood.

Effects of your vehicle

As a competent PCV driver you must always be aware of the effects your vehicle and your driving have on other road users.

You must recognise the effects of turbulence or buffeting your vehicle causes when overtaking

- pedestrians
- horse riders
- cyclists
- motorcyclists
- cars
- cars towing caravans
- other buses and lorries.

Smaller, lighter vehicles are also affected when they overtake you at speed, especially on motorways.

On congested roads, particularly in shopping areas, take extra care when you need to drive close to the kerb. Be aware of

- pedestrians stepping off the kerb
- the danger of your nearside mirror striking the head of a pedestrian standing at the edge of the kerb
- cyclists moving up on the nearside of your vehicle in slow-moving traffic.

16

GETTING STARTED

Passengers

Your job is to deliver your passengers to their destinations

- safely
- on time
- efficiently
- courteously.

Caring for your passengers is as important a part of PCV driving as are the individual driving skills. As the driver, you're responsible for your passengers. Remember, you are the representative of the company. How well you perform this role is a measure of your professionalism.

Many companies have rules governing standards of behaviour required both of you and of your passengers. These are in addition to the more general statutory laws that drivers of PCVs must obey. Make sure that you know the rules and enforce them when necessary.

At some point you'll also find yourself driving to a timetable. This can exert a pressure on you to rush. Resist the temptation to hurry and don't become impatient.

Customer care

- Be on the lookout for passengers. Those waiting might not be able to see or hear the bus coming.
- Eliminate gaps from the kerb. Many passengers find it difficult to board or get off the bus if it pulls up too far away from the kerb. Stop well in to the kerb to help them.
- Look directly at each passenger when you speak. It may make a world of difference to some of your customers.
- Give passengers time to get seated before you move off. A few extra seconds here add very little to journey times but demonstrates good customer care.

GETTING STARTED

Buses taking children to school

Many bus or coach companies throughout the country are responsible for transporting children to and from school during term time. Drivers have often complained about the stress caused by the children's behaviour and the responsibility of having up to 80 children on the bus at one time. Many children consider the driver to be a 'miserable' person who moans all the time and think that the buses are overcrowded and of poor quality.

Bus companies are looking into ways of improving the situation for both drivers and children. Research has been undertaken to find the best way to approach this problem and D*f*T (the Department for Transport) has created a training programme for bus drivers. Copies of the report *The School Run – a Training Programme for Bus Drivers* are available from

D*f*T
PO Box 236
Wetherby
West Yorkshire
LS23 7NB

Tel: 0870 1226 236
Fax: 0870 1226 237

The report includes problems from the drivers' point of view. Many thought that

- children's behaviour had deteriorated and they were unmanageable
- children did not show drivers any respect.

Pupils' views were also highlighted in the report; for example, they thought that

- a degree of high spirits was acceptable
- drivers had forgotten what it was like to be young.

However, pupils also recognised that high spirits could easily get out of hand and lead to unacceptable behaviour.

The training programme suggests ways to make the situation more tolerable and less stressful for all concerned. Effort is needed on both sides for the working relationship between driver and pupils to be successful.

Difficulties encountered on buses taking children to and from school may be dealt with in different ways. Bus and coach companies will have guidance and advice set out by their management teams that should be followed. Examples of good practice include

- consider the safety of yourself and other bus and highway users: the safety of schoolchildren and other members of the public must be your highest priority
- avoid physical contact with schoolchildren under any circumstances other than
  - genuine self-defence
  - a medical emergency
  - to prevent a serious offence or threat to safety
- schoolchildren may not, in words or actions, be told to get off the bus or refused entry
- racist or other offensive or abusive language will not be tolerated or permitted from any driver
- do not make any threats: instead give warnings along the lines of *The School Children Bus Contract*
- drivers should not react to bell-ringing or verbal abuse.

GETTING STARTED

If there is a risk of a disruptive young person or young people causing damage to the vehicle or endangering the safety of themselves, other passengers or you, the driver, you should

- bring the bus to a halt
- ask them to calm down
- if there is no response, read out loud from the card issued to all drivers for use in such instances
- if the disruptive behaviour continues, call for help.

It is essential that you remain calm during any situation and avoid doing or saying anything that implies you are asking them to leave the bus. Act in a confident manner and keep your behaviour in proportion to the provocation. You should

- think about your approach
- minimise the 'audience effect': young people find it harder to back down if they are being watched by their peer group
- be aware of warning signs and think ahead
- show that you are willing to listen
- avoid body language that could inflame the situation, such as pointing of fingers
- inform the school and your management of any persistent problems.

It has been found that using a dedicated driver for a particular school run enables a long-term relationship to develop between the driver and passengers. This helps reduce the need for young people to test the boundaries of what is acceptable.

Professional service

Operators often publicise journeys as being

- comfortable
- convenient
- fast
- trouble-free.

Courtesy and consideration are the hallmarks of a professional driver. Both you and your company, not to mention your profession, will be on display every time you drive. Therefore, you should show a good example of skill, courtesy and tolerance. Be a credit to yourself, your company and your profession, and aim for the highest standards.

Commercial pressure

There's a lot of competition among operators for passengers. Such competition helps to ensure that high-quality services are available. However, competition also means that operators need to have tight cost controls to ensure efficient and effective use of their resources. But cutting corners on safety isn't acceptable and could be a recipe for disaster. Remember, safety must be your first priority.

- You must not drive a vehicle with a serious defect.
- If you're delayed, do your best to make up time, but don't speed or take risks.
- Obstructing or 'racing' another operator's vehicle is inexcusable.

GETTING STARTED

Passengers with mobility difficulties

'If someone smiles and takes the money with a little bit of patience it makes the world of difference.'

'Just speaking carefully, looking at the person and giving them attention – not feeling rushed – matters a lot. The feeling that you're holding up a queue of people is a very anxiety-producing situation.'

These are comments from passengers about their local bus service. They're the sort of people you might carry every day – regular customers, in fact. Yet both of them have a problem that may be hard to recognise: they're disabled.

Some disabilities are very obvious. A person carrying a white stick, a long white cane or accompanied by a guide dog is visually impaired. If the stick has a red ring painted on it they also have impaired hearing. It's also easy to see that someone with crutches, a walking frame or any other aid to movement has a disability – perhaps only temporarily.

Remember, showing a little consideration goes a long way with most people – whether they have special needs or not.

Try to imagine what assistance you would like if you were in the position of a person with mobility difficulties.

GETTING STARTED

Be patient and considerate. Always respect their wishes: disabled people want to retain their independence. If someone tells you they can manage – let them. But be prepared to offer help if they appear to need it, or ask for it. You'll have your own problems to cope with – such as trying to keep to time, busy traffic conditions, inconsiderate behaviour by other road users – but you should do your best to offer courtesy and a smooth ride to those with special needs.

Also, think about the everyday problems faced by people trying to manage with children, pushchairs or shopping trolleys. Allow time for pushchairs to be stowed away in the correct place. Make sure they are not left in a place that would endanger other passengers attempting to get past. They should be correctly stowed to prevent them being thrown forward in the event of an accident.

Blind and partially sighted people

More than 200,000 people in the UK are visually impaired. Only a small proportion are totally blind, but you may not be able to tell by their appearance. Visually impaired people often depend on their local bus service for mobility. And remember, most partially sighted people find it hard to read a destination display or timetable.

'There's a problem of explaining that we can't see very well ... we want to do as much as we can for ourselves and just be helped with the tiny bit that we can't do ...'

People who are deaf or hard of hearing

'I usually ask the fare and, if I don't hear how much it is, sometimes I bluff and just offer £1 and hope to get the right change. If the bus driver seems to be a pleasant, approachable person I don't mind asking him to repeat it, but some drivers are under pressure and appear not to be aware of you or don't look at your face ...'

It's common courtesy to look at people when you speak to them. Just doing that will allow most deaf or hearing-impaired people to understand you. Good communication also saves time.

GETTING STARTED

Physical disabilities

People with arthritis, stiff joints, artificial limbs or conditions such as multiple sclerosis, often put up with extra pain (and the impatience of other passengers) rather than ask for extra consideration on a bus. For them, courtesy and a smooth ride are important.

'Nobody wants to shout to the rest of the world "I am having trouble", but if the driver could just wait until you're sitting down before they pulled away ...'

'If letting the clutch out or moving away is done too violently it hurts every inch of the way ...'

'If the driver was to go round corners a little more slowly it would probably be less painful ...'

Lifts, ramps and 'kneeling' buses

Make sure that you're thoroughly trained in the safe use of passenger lifts, ramps and securing devices. If you drive a vehicle fitted with this equipment, never let untrained people operate it. Watch out for the safety of others at all times.

Some buses are equipped with air or hydraulic systems that allow the step level to be raised and lowered. These 'kneeling' buses improve access for disabled and elderly passengers. It's essential that you're thoroughly trained in the use of such systems and are aware of the principles of safe operation.

Learning disabilities

Customers with learning disabilities may appear fit and active, but they may also find bus travel a special problem and a challenge. It may be hard for them to understand other people or to make themselves understood. Also, any unexpected problems can sometimes produce a sense of panic.

Those with learning disabilities are increasingly being encouraged to go out to work, to go shopping or visit friends. With patience and understanding you can contribute towards their confidence and sense of achievement.

GETTING STARTED

Part Two

Passenger carrying vehicles

The topics covered

- **Forces at work**
- **Maintaining control**
- **Vehicle sympathy**
- **Types of PCV**
- **Vehicle maintenance**
- **Other auxiliary systems**

PASSENGER CARRYING VEHICLES

Forces at work

You should understand something of the various forces that act on a vehicle and its passengers. The effects of these forces can seriously undermine your control, so it's important to be aware of them and to act appropriately.

A bus, coach or minibus travelling in a straight line under gentle acceleration is relatively stable.

When a vehicle

- accelerates
- brakes
- changes direction

forces are applied to it and its load. The more violent or sudden the change, the greater the forces. Sudden, excessive or badly timed steering, braking and acceleration will introduce forces that can result in loss of control.

Steering should always be

- planned
- smooth
- controlled
- accurate.

Braking should always be

- progressive
- correctly timed
- smooth
- sensitive.

Acceleration should always be

- progressive
- used to best economic advantage
- well planned
- considerate.

Most of the forces described here act on a vehicle in motion. If you disregard them you'll probably lose control, so allow for them in your driving.

PASSENGER CARRYING VEHICLES

Friction

The resistance between two surfaces rubbing together is called friction. A tyre's grip on a road surface depends on friction, and is essential when

- moving away or accelerating
- turning/changing direction
- braking/slowing down.

The amount of grip will depend on

- the weight of the vehicle
- the vehicle's speed
- the condition of the tyre tread
- the tyre pressure
- the type and condition of the road surface
    - loose
    - smooth
    - anti-skid
- weather conditions
- any other material present on the road
    - mud
    - wet leaves
    - diesel spillage
    - other slippery spillages
    - inset metal rails
- whether the vehicle is braking or steering sharply
- the condition of steering and suspension components.

Sudden acceleration or braking can lead to loss of grip between the tyre tread and the road surface. Under these conditions the vehicle may

- lose traction (wheelspin)
- break away on a turn (skid)
- not stop safely (skid)
- overturn.

The same will happen when changing into a lower gear if travelling too fast or if the clutch is suddenly released, because the braking effect will only be applied to the driven wheels.

26

PASSENGER CARRYING VEHICLES

Gravity

When a vehicle is stationary on level ground the only force acting upon it is the downward pull of gravity (ignoring wind forces, etc.). On an uphill gradient the effects of gravity will be much greater so that

- more engine power is needed to move the vehicle forward and upward
- less braking effort is needed and the vehicle will stop in a shorter distance.

On a downhill gradient the effects of gravity will tend to

- make the vehicle's speed increase
- require more braking effort
- increase stopping distances.

The vehicle's centre of gravity is the point around which all its weight is balanced. All passenger vehicles are 'tilt tested' to ensure that the design is stable. But violent steering, acceleration or braking shifts the centre of gravity and places excessive forces on the vehicle's tyres and suspension, and on the passengers.

Heavy braking whilst cornering can bring components very close to their design limits and will be uncomfortable for passengers. Catching a kerb or raised drain cover with a tyre under such conditions could result in a blow-out and the vehicle going out of control or even overturning.

Centrifugal force

When a vehicle takes a curved path at a bend the forces acting upon it tend to cause it to continue on the original, straight course. This is known as centrifugal force. If a bus or coach takes a bend too fast centrifugal force will cause the passengers to be thrown towards the outside of the bend. The vehicle may even skid, especially if the road surface is at all slippery.

27

PASSENGER CARRYING VEHICLES

Inertia and momentum

A stationary bus with 70 or 80 passengers on board may weigh up to 18 tonnes. It requires a great deal of force to begin to move it, even on a flat road, but it takes relatively little power to keep it rolling at a constant speed. Resistance to change in a vehicle's state of motion is called inertia, and the force that keeps the vehicle rolling is called momentum.

Modern buses and coaches have engines with a high power output to

- give good acceleration
- overcome inertia.

Passengers are also affected by these forces. A passenger's inertia has to be overcome in much the same way as the vehicle's. Acceleration will push them back into their seats, while braking will move their weight forward, due to momentum. Sudden braking will cause passengers to be thrown forward and could be dangerous. Therefore, all acceleration and braking should be smooth, controlled and as progressive as possible.

Kinetic energy

The energy that's stored up in the vehicle and its passengers when travelling is known as kinetic energy. This is converted into heat at the brake shoes and drums when braking occurs.

Continuous use of the brakes results in them becoming over-heated and losing their effectiveness (especially on long downhill gradients). This effect is known as brake fade.

Much more effort is needed to stop a fully-laden PCV than an ordinary car travelling at a similar speed. It's important, therefore, to avoid harsh braking. Plan ahead and take early action.

PASSENGER CARRYING VEHICLES

Maintaining control

You can't alter the severity of a bend or change the weight of the bus and its passengers. Similarly, you can't alter the design and performance characteristics of your vehicle and its components. But you do have control over the speed of your vehicle and hence the forces acting upon it.

If you ask too much of your tyres by turning and braking at the same time, you'll lose some of the available power and grip. When the tyres slide or lift you'll no longer be in full control of the vehicle. To keep control you should ensure that all braking is

- controlled
- in good time
- made when travelling in a straight line, wherever possible.

Reduce speed in good time by braking, if necessary, before negotiating

- bends
- roundabouts
- corners.

Avoid braking and turning at the same time, unless manoeuvring at low speed. Reduce your speed first and look well ahead to assess and plan.

PASSENGER CARRYING VEHICLES

Vehicle sympathy

There are many different types of PCVs and each type will require specific handling. Adapt your driving to suit the vehicle and develop what is known as 'vehicle sympathy'.

For example, drivers need to take corners slowly in order to keep their passengers comfortable. Yet it's difficult to define what 'slowly' means for all vehicles on all occasions. A safe, comfortable speed will depend on the sharpness of the corner and any other hazards there might be. The vehicle's design might dictate when the speed is comfortable. New coaches have very sophisticated air-suspension levelling systems, which allow relatively fast cornering whilst maintaining the body almost level.

The implementation of the Disability Discrimination Act means more buses and coaches will have wheelchair users travelling on them, so their comfort must be considered.

Information at the back of this book gives details on the type of licence needed to drive the different types of PCVs. This section discusses some of the basic characteristics of the various types of PCVs. However, it's up to you to develop your own 'vehicle sympathy' when driving.

PASSENGER CARRYING VEHICLES

Types of PCV

Minibuses

A minibus is generally defined as a motor vehicle with more than eight and not more than 16 passenger seats. They are often based on van bodies and have been adapted by specialist coach-building firms, although some manufacturers produce purpose-built vehicles. The controls are usually similar to those on cars.

Few minibuses are built as full public service vehicles. The regulations for PSVs require higher minimum standards for items such as

- headroom
- access
- seating
- safety precautions
- equipment
- markings.

Driving minibuses

Driving a minibus is often a lot like driving a car. However, you need to be aware that despite power-assisted steering and braking, and possibly an automatic gearbox, it can be more demanding and tiring than driving a car.

You're strongly advised to seek professional training if you intend to drive minibuses. Various bodies run courses, but if you have difficulty finding one locally, contact RoSPA, whose address and telephone number are at the back of this book.

Information at the back of this book tells you about licence requirements dependent on usage. Those vehicles operated under a community or minibus permit scheme are subject to special rules.

When driving a minibus you'll need to think about the following

- weight
  - greater stopping distances are needed
  - they're slower to accelerate and to overtake
  - more effort is needed for steering.
- height
  - there's greater body roll, pitch and sway
  - they're more susceptible to side winds, etc.
- noise levels
  - these can be high, especially in van-derived models
  - passenger noise can be high and distracting.
- speeds
  - it's more difficult to maintain high average speeds
  - when fully laden, speed may be lost rapidly on uphill stretches of road.
- passengers' comfort.
- distances travelled
  - is the vehicle suitable for long journeys?
  - would the use of a larger vehicle, possibly hired with a driver, be more appropriate?

31

PASSENGER CARRYING VEHICLES

- Time
  - plan your journey and estimate realistically how long it will take
  - allow plenty of time for the journey, thus putting yourself under less pressure
  - you'll need to take adequate breaks.

Never drive for more than four and a half hours without taking a break of at least 45 minutes. If you're subject to drivers' hours regulations, you'll find that this rule, and others, are legal requirements. To avoid fatigue it is advisable to have a break after two hours' driving.

Treat minibus driving as you would other work, even if it isn't your normal job. You need to be alert and to concentrate. Refer to the rules in Part Three, which apply to professional drivers, and consider the advice in the officially recommended syllabus in Part Five.

Ultimately, consider carefully before each journey whether

- you need someone else to drive
- a second driver is advisable.

Seat belts

Seat belts save lives and reduce the risk of injury. Current legislation requires that, when three or more children aged between three and 15 years (inclusive) are carried on an organised outing in a minibus, larger minibus or coach, they must be provided with, as a minimum, a forward-facing seat fitted with a lap belt. An organised outing includes the school run, even when driven by parents. Minibuses include less obvious vehicles such as large domestic vehicles that have more than eight and not more than 16 seated passengers.

In all buses, coaches and minibuses, the driver and all front seat passengers must wear a seat belt, where it is fitted.

If the unladen vehicle weight is 2,540 kg or less, and belts are fitted, then all rear-seat passengers are obliged to wear them. In law, the driver is legally liable if children aged under 14 years are not restrained.

- Children under three years old in the front must use an appropriate child restraint (i.e. a baby or child restraint). In the rear, an appropriate restraint must be used if one is available – if not, then the adult belt must be used. Do not permit an adult to put one seat belt around both themselves and an infant on their lap. That could result in severe injuries to the child in the event of a crash.

- Children aged between three and 11 years and under 1.5 metres in height, in the front or rear, must use an appropriate restraint if one is available; if not, then an adult seat belt must be used.

- Children over 1.5 metres in height, in the front or back, must wear an adult seat belt.

In buses and coaches, and in minibuses over 2,540 kg it is recommended that all other passengers in any seats other than front seats wear a seat belt where it is fitted.

The law governing the wearing of seat belts by rear seat passengers is currently under review.

PASSENGER CARRYING VEHICLES

Midibuses

There's no legal definition of a midibus. However, the term is commonly used within the industry to describe any single-deck vehicle that's between a minibus and a 40+ seat coach or bus.

Virtually all are purpose-built and many have bus or coach controls, equipment and other systems. Some midibuses are specialist vehicles with wheelchair lifts and securing equipment. Many are used on normal services, where demand isn't sufficient to justify the use of full-size buses.

Seat belts

Refer to information given in the seat belt section on page 32.

Depending on the use and seating capacity, drivers require one of the following licence entitlements

- D
- D1
- D + E or D1 + E, if a trailer over 750 kg is to be towed.

See pages 239 and 242 for more information on licence entitlements and requirements.

Some midibuses operated under the community minibus and large bus permit schemes can be driven with a category B (car) licence. The rules are explained in booklet PSV 385, available from Traffic Area Offices.

It's essential that you fully understand the vehicle controls and, wherever possible, undergo 'type' training.

Many of the points relating to minibuses also apply to midibuses, as do many of the topics covered in the sections on buses and coaches. In particular, you'll need to consider

- blind spots and restricted vision
- standing passengers
- careful use of automatic gearboxes, where fitted
- body roll.

PASSENGER CARRYING VEHICLES

Single-deck service buses

These vehicles are generally designed for local bus service use and have basic passenger equipment. They may also have a limited amount of seating and a higher proportion of space for standing passengers. Most are one-person operated.

Newer vehicles are built to the Disabled Persons Transport Advisory Committee (DPTAC) specification and may incorporate 'kneeling' suspension, wide doors and other design features to cater for passengers with disabilities. Because of the 'stop–start' nature of the journeys, most of these vehicles have semi-automatic or fully automatic gearboxes, although some buses with manual gearboxes are still in use. All have relatively low gearing, with only four or five gears, or are coupled to low-ratio drive axles to give greater flexibility at low speeds. As a result they may have lower top speeds.

Single-deck service buses require skill and sensitivity on the part of the driver if they're to be driven smoothly.

While these vehicles and double-deck buses are generally exempt from the requirement to fit seat belts, be aware that some contracts, particularly those with schools, may require that vehicles fitted with seat belts shall be provided for passengers.

See page 32 for more information on seat belt requirements.

PASSENGER CARRYING VEHICLES

Double-deck service buses

These are high-capacity vehicles used primarily for local bus services. Many are fitted with dual doors and fare-collection equipment to allow for effective one-person operation.

Additional internal mirrors are positioned to allow the driver to observe entrances, exits, stairs and the upper deck. To ensure high standards of passenger care and safety, drivers should make full use of these mirrors.

Automatic and semi-automatic gearboxes are frequently fitted to these vehicles. Make sure that you know how to make smooth gear changes and to use the gearbox correctly when moving off and pulling up. Vehicle manufacturers give advice for each type of vehicle.

Drivers need to balance safe driving techniques with the comfort of passengers and the need to keep to timetables. Smooth, skilful driving will be essential during peak periods when there will be more passengers standing, climbing the stairs and moving about the bus.

Most modern double-deck vehicles have underfloor or rear-mounted engines. You're less likely to know if the engine is overheating, for example, so you'll have to make full use of instruments and warning lights to ensure early action should a fault develop.

On double-deck buses the driving position and front entrance are generally ahead of the front axle, whilst the position of rear axles varies considerably. The wheel-base of the bus will dictate the appropriate course to take when cornering. This means that you must take care with overhangs and be aware of the danger of tyre damage on kerbs etc.

Read the information on vehicle height in Part Three and take extra care when driving open-top double-deck buses, such as 'sightseeing' tour buses and those operated in seaside towns.

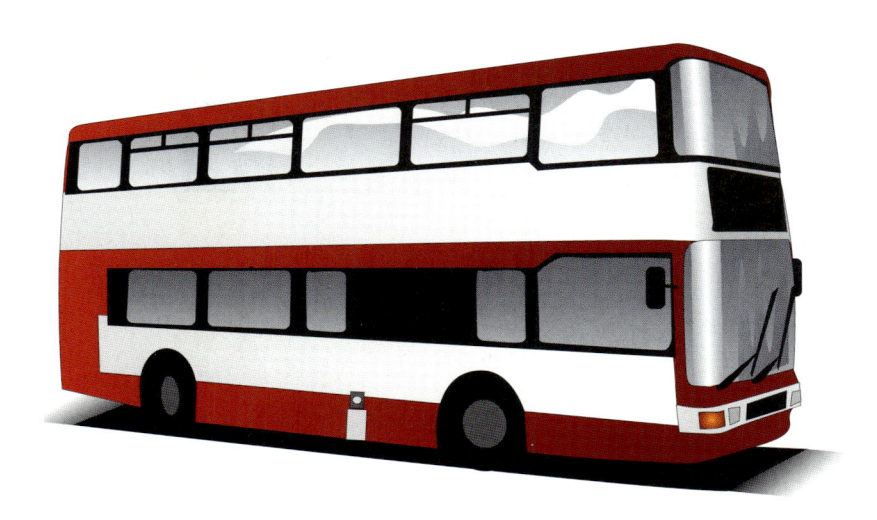

PASSENGER CARRYING VEHICLES

Articulated buses

Articulated buses, also known as 'bendi-buses', consist of a two-axle lead unit coupled to a single-axle rear section by means of floor- and roof-level pivots and a flexible shroud. They offer high-capacity urban transport on routes where double-deck buses are less practical.

Trials with articulated buses have taken place in a number of areas in the UK, and their numbers are increasing. They are more common in other countries, particularly in Europe, where height limits of 4 metres (13 feet) exist. In this country their length – up to 18 metres long (59 feet) – can present problems when used on urban streets.

Additional care is needed when driving these vehicles. Always be aware of the 'swept path' the vehicle is taking. And remember, the rear section, unlike some large articulated goods vehicles, exactly 'tracks' the path taken by the lead section. 'Type' training is essential before driving an articulated bus.

When negotiating road junctions and pulling into lay-bys, you must remember to make allowances for the additional length of the vehicle. Make sure you don't obstruct other road users.

Try to avoid getting into situations where you need to reverse the vehicle. Only reverse when special video reversing equipment is fitted or a reliable person is standing in your view to guide you back.

New drivers of vehicles towing trailers will need to take a category D + E test if the trailer is over a maximum authorised mass of 750 kg. An articulated bus isn't deemed to be a bus towing a trailer and can therefore be driven on a category D licence.

PASSENGER CARRYING VEHICLES

Single-deck coaches

Coaches are designed to carry passengers for longer distances, in greater comfort and with improved facilities. Many have sophisticated heating and air conditioning systems toilets, catering areas and crew seats. Most modern vehicles are fitted with rear or underfloor engines, to limit noise levels and to enable more luggage to be carried.

If vehicles are fitted with video and television equipment for passenger use, it's illegal for their screens to be visible to the driver while they are in use.

Special regulations apply to the charging, use, location and emptying of water and toilet systems fitted to road vehicles. See the relevant advice in the officially recommended syllabus in Part Five.

Coach journeys are longer and frequently use motorways, so manual gearboxes remain the norm, but they often have six or more gears. Semi-automatic and fully automatic vehicles are also in use, and there's an increasing trend towards air suspension systems.

Coaches are often downgraded to dual-purpose or service-bus use after several years of operation. In addition, some rural bus operators use coaches so that their passengers travel in greater comfort. In such instances, lower specification running gear may be fitted to the vehicles to make it easier for the driver (less gear-changing, etc).

For seat belt information, see the seat belt section on page 32.

PASSENGER CARRYING VEHICLES

Double-deck coaches

The first double-deck coaches were introduced in the UK in the 1950s. They were based on bus body shells, but were fitted with more powerful engines and higher gearing. Coach seats were added to provide high-capacity, luxury vehicles able to compete with other long-distance passenger transport.

Since then there have been considerable developments, not least in the facilities double-deck coaches now offer. Nearly all are now specially designed and purpose-built. Comfort and customer service are the biggest selling points. Although some double-deck coaches provide 70 or more seats, passenger-carrying capacity isn't always the key attraction to customers. These coaches may be fitted with

- toilets
- refreshment facilities
- lounges
- tables
- telephones and fax machines
- audio visual equipment
- crew sleeping accommodation.

A number of double-deck coaches have a courier service and some specialist vehicles are designed to carry as few as 12 passengers, with full sleeping or conference facilities provided.

The regulations governing video and television equipment and waste water disposal are similar to those for single-deck coaches: these are covered in more detail in the syllabus in Part Five.

These coaches are amongst the most sophisticated vehicles on the road, with high-power engines, versatile manual, automatic, semi-automatic or electronic gearboxes, air suspension and power-assisted controls. Make sure that you understand all the systems fitted to the vehicle and are fully competent to operate them.

For seat belt information, see page 32.

Driving positions may be unusual in these vehicles, so

- the driver may not be able to see what's happening inside the coach
- video or electronic sensor systems may be fitted to help with manoeuvring and to add to the view given by the rear view mirrors
- additional mirrors may be fitted to show the driver what's happening below his or her field of vision at the front of the coach.

Use all these aids when driving to help you to drive safely.

PASSENGER CARRYING VEHICLES

Tri-axle buses and coaches

Higher vehicle weight has meant that air suspension is being fitted increasingly to all but the lightest PCVs to improve passenger comfort. It helps counter the damaging effects on roads and bridges and assists the dynamic handling of the vehicle. Another recent development has been the addition of an extra rear axle to further distribute vehicle loads.

Handling isn't greatly different from two-axle vehicles, except that punctures and blow-outs are sometimes difficult to detect. Frequent tyre checks are advised.

The course the wheels take on tight corners should be observed and allowed for when driving. Very low speed is advisable when the steering is on full lock to minimise any possible 'scrubbing' effect on the rearmost tyres.

Follow the seat belt instructions given on page 32 if your tri-axle coach is fitted with seat belts.

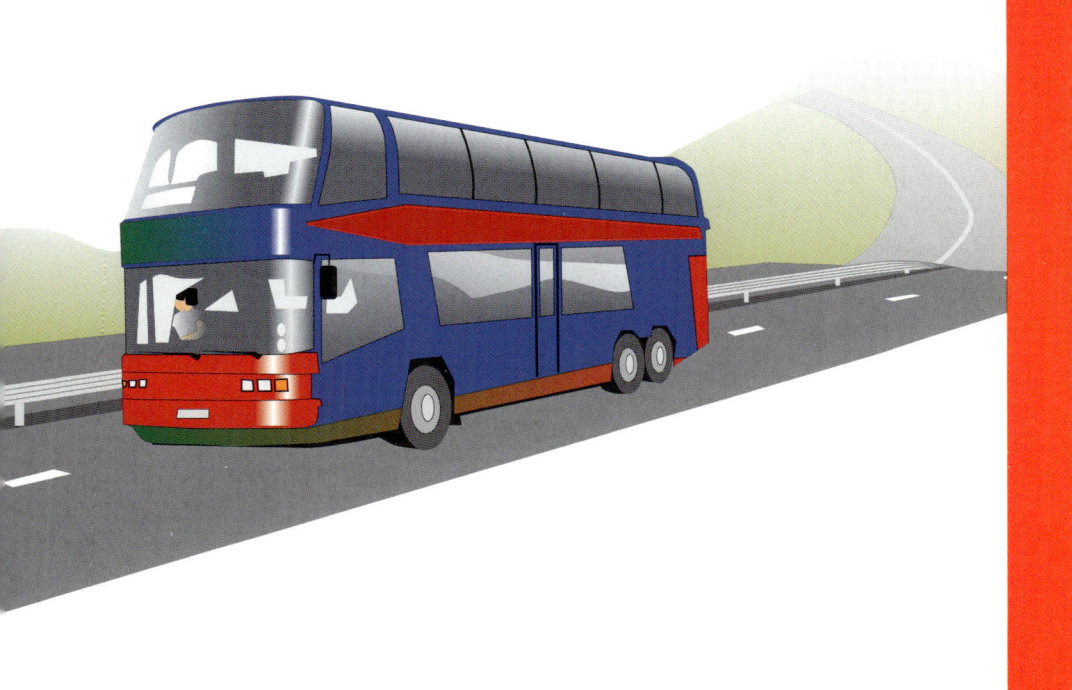

PASSENGER CARRYING VEHICLES

Mobile project and playbuses

More than 500 double- and single-decker buses and coaches have been converted for community use in the UK. As their primary purpose is for recreational, vocational or educational use, they aren't regarded as PCVs.

There are particular rules for their use and licensing requirements (see Part Seven). They may, in some cases, be driven by category B (car) licence-holders. However, the driving requirements for these large vehicles are the same whether an additional driving test has to be taken or not. If you drive one of these vehicles, it's essential that you're fully aware of your responsibilities.

This book tells you what's expected of professional PCV drivers, but the advice applies to anyone who drives buses or coaches. A book can teach you the basic facts and theory about driving, but you should always seek professional guidance before driving on public roads. You can't expect to drive a bus, whatever its use, without adequate training.

Most mobile project and playbuses are elderly buses that are 'life expired' for PCV operations. The importance of safety checks and adequate maintenance is greater as a result. Drivers must be able to identify faults and understand procedures for putting them right.

Operators and drivers of mobile project and playbuses need to consider the safe

- stowage of equipment when the bus is being driven
- manoeuvring of the bus when arriving at, or departing from, sites
- installation and stowage of any heating, lighting or cooking equipment, including gas cylinders
- operation of generators and fuel storage.

Detailed guidance is available from the National Playbus Association, whose contact details are given in Part Seven.

PASSENGER CARRYING VEHICLES

Historic buses and coaches

Enthusiasts have ensured that many historic buses and coaches have been preserved and are shown at rallies.

Some of these historic vehicles may be driven on a category B (car) licence provided certain rules are observed. These are

- the driver must be over 21
- the vehicle must carry no more than eight passengers plus the driver
- for non-commercial use only

Drivers with category D entitlement may drive historic buses and coaches as they would any other PCV.

You should seek professional training if you intend to drive historic buses and coaches. For example, you need to know how to 'double de-clutch' (refer to the Glossary of terms) or 'snatch-change' to use crash or part-synchromesh gearboxes. These are special techniques that you should practise after they've been explained and demonstrated to you. Also, if the vehicle you drive has air or vacuum brakes, make sure that you understand the meaning of any warning signals.

When you drive a historic vehicle for the first time, start by mastering steering, gear-changing and braking techniques

- off the road
- under supervision
- without passengers.

Older buses and coaches are more difficult to drive than modern counterparts. Generally, there's no power steering, air-assisted clutches or semi-automatic gearboxes to make driving easier.

When driving these older vehicles

- think how your slower speed affects other road users
- pull over to let others pass, when you can do so safely
- treat the vehicle with respect and ask for advice if you come across controls or warning systems that are unfamiliar
- make sure that you have full control.

It's important that you never drive a preserved vehicle unless you're certain that it's fully roadworthy. Carry out all the checks advised in Parts Two and Three and also make sure that you're competent to drive the vehicle.

PASSENGER CARRYING VEHICLES

Passenger and general safety

Never drive a bus in which you have no contact with passengers, without one designated, responsible person in charge of the passenger saloon(s). The exceptions to this are when no passengers are carried and when access to the vehicle is prevented by means of a door, chain, strap or other barrier. In addition, never

- allow passengers to board or alight from the vehicle, unless the parking brake is applied
- allow passengers to ride on open platforms or with open doors
- operate the doors whilst the vehicle is in motion
- allow more passengers to be carried than the vehicle is designed for, or the law allows
- allow bells to be used other than in the accepted way. In half-cab vehicles this is the only means of communication between the passenger saloon(s) and the driver.

The use of conductors has diminished and now they may only be found in London on the Routemaster buses. If you are required to use bell codes, ensure they are understood. As a driver, be aware that passengers may use the bell incorrectly. The bell codes are

- one bell – stop when safe
- two bells – move off when safe
- three bells – bus full
- four bells – emergency on bus.

Always take great care on rally sites when pedestrians are close to moving vehicles. Drive only at walking pace, or slower, and use marshals or other responsible people to help you to manoeuvre safely.

PASSENGER CARRYING VEHICLES

Light rail (or rapid) transit (LRT) systems

Trams are often referred to in several ways. They may be called light rapid transit (LRT) or 'metro' systems, or 'supertrams'. LRT systems are common throughout Europe and there are plans to introduce them to many more cities in the UK.

They're essentially modern tramways – the vehicles running singly or, more often, as multiple units on standard railway track gauge to light railway specifications. Some systems operate completely segregated from other traffic and may run on former railway tracks. LRT vehicles are fixed in the route they follow and can't manoeuvre around other vehicles and pedestrians.

The area occupied by an LRT vehicle is marked by paving or markings on the road surface. This 'swept path' must always be kept clear. Other road users, including bus and coach drivers, must avoid blocking 'supertram' routes.

The following points are important.

- In some towns and cities certain roads are restricted to buses and LRTs only.
- Where LRTs operate on roads not segregated from other traffic, LRT drivers must hold full category B licence entitlement.

- LRT drivers and vehicles are subject to all the normal rules of the road, in addition to specific rules about LRT operation.

Drivers are only permitted to operate 'supertrams' after extensive training. All UK LRT and traditional tram operators have dedicated training schools and staff to ensure high safety standards.

Other road users need to be aware of how to deal with LRTs – and of their limitations. When a tram approaches, other vehicles (and pedestrians) must

- keep away from the swept path area
- obey yellow box junction rules and not block junctions
- anticipate well ahead and never stop on or across the tracks
- obey all traffic light signals and never 'jump' lights that show the tram has priority.

Open-top buses shouldn't be driven beneath overhead LRT power supply lines. Also, whenever possible, drivers of non-tram vehicles should avoid driving directly along metal rails, especially in wet weather, to avoid the risk of skidding.

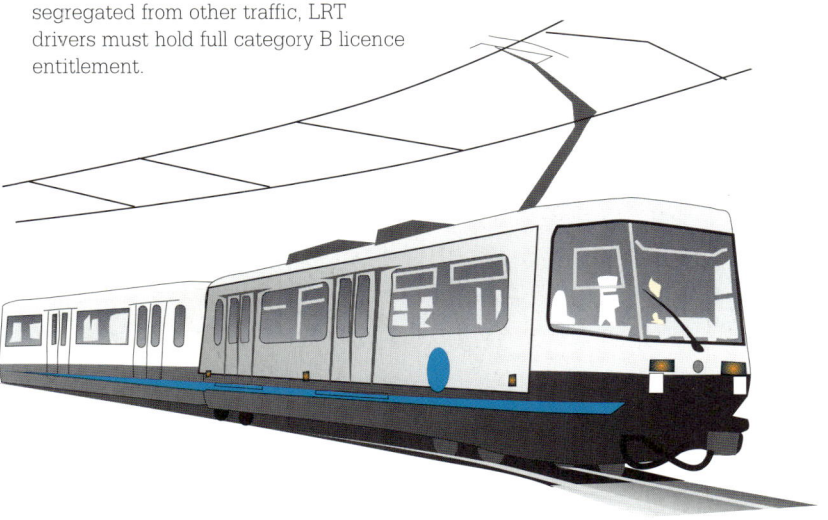

PASSENGER CARRYING VEHICLES

Tram signs

Warning of trams crossing ahead

Speed limit for tram drivers (all diamond-shaped signs are only for tram drivers)

Trams travel in both directions All other traffic obeys one-way signs

The signal mounted to the right gives instructions to tram drivers, which may not be the same as those given to drivers of other vehicles

Reminder to pedestrians to look out for trams approaching from both directions

Lane for trams only

Route for trams only

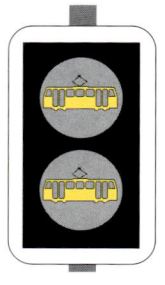

Warning signals for pedestrians – the lights flash when a tram is approaching

44

PASSENGER CARRYING VEHICLES

Towing trailers

Considerable care is needed when towing a trailer, especially when reversing. Extensive training and practice are strongly recommended.

When you tow a trailer make sure that

- access to emergency exits aren't obstructed
- you know and comply with the speed limits that apply to vehicles towing trailers
- you don't carry passengers in the trailer.

Any unattended trailer is a road hazard, especially at night or in poor visibility, such as foggy conditions.

New EC regulations are now in force covering the towing of trailers by motor vehicles. The information in Part Seven details how this will affect drivers of PCVs.

When uncoupling a trailer, select a suitable site. It should be safe and on firm and level ground. Make sure you apply the trailer handbrake before commencing the uncoupling procedure (see page 223).

45

PASSENGER CARRYING VEHICLES

Vehicle maintenance

Preventative maintenance

It is important to keep your vehicle well maintained; breaking down whilst on the road can have road safety implications. Follow manufacturers' guidelines for service intervals. In addition to this, being aware of components wearing out or requiring replacement will help prevent costly breakdowns for your company. Neglecting the maintenance of vital controls and fluids such as brakes, steering and lubricants is dangerous; they need to be checked regularly.

Ensuring that the daily walk-round checks are carried out will enable you to find any defects that could become a problem and cause the vehicle to break down or be driven whilst illegal. The time taken to complete a thorough check will be less than that required to organise repair or replacement whilst out on the road.

Checks need to be made before you start up the vehicle or begin a journey. The consequences are too great to risk driving a vehicle with defective parts.

Daily checks

You need to check the following regularly to ensure your vehicle is well maintained and not in need of attention. Check

- there are no fuel or oil leaks
- the security and condition of your battery
- tyres and wheel fixings
- spray suppression equipment (if fitted)
- steering
- excessive engine exhaust smoke
- brake hoses
- coupling security (if applicable)
- brakes.

See also the daily walk-round checks on page 77; they will help you to notice if any part of your vehicle needs maintenance. Always refer to the handbook for your individual vehicle before carrying out any maintenance tasks and follow any safety guidance it may contain.

Technical support

Traffic Commissioners and the Vehicle and Operator Services Agency (VOSA – formerly the Vehicle Inspectorate and Transport Area Network) will provide advice and assistance to operators on safety inspection intervals. VOSA offers a brake performance check service, headlight alignment and an emission check at all of its full-time heavy goods vehicle testing stations.

Construction and functioning of the internal combustion engine

There are two main types of internal combustion engine

- spark ignition (petrol) – the fuel and air mixture is ignited by a spark
- compression ignition (diesel) – the rise in temperature and pressure during compression causes spontaneous ignition of the fuel and air mixture.

During each revolution of the crankshaft there are two strokes of the piston: the piston travels both up and down the engine cylinder. Both types of engine can be designed to operate using a two-stroke or four-stroke principle. Almost all modern passenger carrying vehicles use the four-stroke principle.

PASSENGER CARRYING VEHICLES

The four-stroke operating cycle

- **Induction stroke** – the open inlet valve enables the piston to draw in a charge of air when travelling down the cylinder. With spark ignition engines the fuel is usually pre-mixed with air.

- **Compression stroke** – both inlet and exhaust valves close and the piston travels up the cylinder. As the piston approaches the top, ignition occurs. Compression ignition engines have the fuel injected towards the end of the compression stroke.

- **Expansion or power stroke** – combustion created throughout the charge raises the pressure and temperature and forces the piston down. At the end of the power stroke the exhaust valve opens.

- **Exhaust stroke** – the exhaust valve remains open, the piston then travels up the cylinder and remaining gases will be expelled. When the valve closes, residual exhaust gases will dilute the next charge.

Diesel fuel system

Compression ignition, commonly called diesel, engines are now almost universally used for large goods vehicles and passenger carrying vehicles.

The fuel injection system functions by delivering a fine spray of a precisely controlled amount of fuel at very high pressure and at the correct time into the engine cylinder combustion chamber. A throttle butterfly valve operated by the accelerator pedal controls the amount of air delivered to the engine and the fuel quantity is adjusted to suit.

Many engines are turbocharged, where the exhaust gas drives a turbine, which compresses the incoming air and effectively delivers more air to the engine. For a given size engine the power is increased and torque is both increased and maintained over a wider engine speed range than the non-turbocharged or normally aspirated engine. Both result in improved vehicle performance.

Never use poor quality diesel fuel. This may lead to increased wear of the injection pump and early blockage of fuel injector nozzles. In winter the composition of diesel fuel is altered by the use of additives to lower the temperature at which waxing or partial solidifying of the fuel occurs. Winter grade fuels should be perfectly satisfactory in all but very severe conditions. Electrically powered fuel line heating systems are available if required.

Open the water drain valve, usually fitted to the base of the fuel filter, at least at the intervals recommended by the vehicle manufacturer.

Bleeding of fuel systems

It may become necessary to bleed the fuel system to remove any trapped air if

- the engine is new or has been renovated
- the fuel system has been cleaned or the filter changed
- the engine has not been run for a long time
- the vehicle has been driven until the fuel tank is empty.

Engine lubrication system

The engine uses a pressure-fed, full-flow, wet sump system. The oil filter, which is normally disposable, contains a bypass valve, which operates if the filter becomes blocked. A pressure relief valve controls the oil pressure; this is housed in the oil pump housing. The oil pump is driven directly from the engine.

Oil is drawn from the sump to the oil pump via a wire mesh pre-filter. The oil circulates from the pump through the main filter, which collects sediment from the oil. The oil then passes to the engine bearings and

PASSENGER CARRYING VEHICLES

other moving parts. Having completed its circle, the oil drains back into the sump.

Always use the recommended type and viscosity of lubricant as suggested by the manufacturer. The oil should also be changed at the required recommended intervals. Friction and wear will reduce the life expectancy and the performance of a vehicle. Friction increases when there is direct metal-to-metal contact between sliding parts. Lubrication helps prevent such contact by reducing wear from friction and heat on working parts within the engine. A film of lubrication covers the various surfaces to keep them apart and maintain fluid friction rather than a dry friction.

Lubrication prevents corrosion of the internal components in the engine. It removes the heat generated in the bearings or caused by combustion and absorbed by metal components. It is also able to seal piston rings and grooves against combustion leakage.

Checking oil levels

You need to check the oil frequently: make sure the vehicle is parked on a level area not on a slope. Check the oil while the engine is cold for a more accurate result. If your vehicle is fitted with automatic transmission there may be an additional dipstick for transmission oil level checks.

You should not run the engine when the oil level is below the minimum mark on the dipstick. Don't add so much oil that the level goes above the maximum level, this creates excess pressures that could damage the engine seals and gaskets and cause oil leaks. Moving internal parts can hit the oil surface in an over-full engine causing possible damage and loss of power.

If the oil pressure warning light on your instrument panel comes on when you're driving, stop and check the oil level as soon as it is safe to do so. If the level is satisfactory, there may be a more serious problem such as failure of the oil pump, which would lead to severe engine damage.

Lubrication oil – engine

The oil in your engine has to perform several tasks at high pressures and temperatures up to 300° C. Lubrication resists wear on moving surfaces and combats the corrosive acids formed as the hydrocarbons in the fuels are burnt in the engine. Engine oil also helps to keep the engine cool. Use the lubricant recommended in the vehicle handbook.

Lubrication oil – gearbox

Most vehicles have a separate lubricating oil supply for the gearbox; it is especially formulated for gearbox use. Follow the instructions in the vehicle handbook.

Engine coolant

It is generally recognised that using an approved coolant solution, containing an anti-freeze additive, throughout the year will give you the best protection. Coolants ensure the cooling system will be protected from freezing in cold weather.

In addition to the anti-freeze agent, coolant contains a corrosion inhibitor, which reduces oxidation and corrosion in the engine and prolongs the life of the cooling system. The anti-freeze additive is an inhibitor called ethylene glycol that has a boiling point of 195° C compared to water at 100° C. The coolant solution is usually diluted with the same volume of water to give maximum protection.

Check the coolant level frequently; if you need to top up regularly it might indicate a leak or other fault in the cooling system that will require checking. Never remove the radiator cap to refill when the engine is hot, always allow the engine to cool before adding further diluted coolant. Don't overfill the system, as the excess will be expelled as soon as the engine warms up.

PASSENGER CARRYING VEHICLES

Transmission system

A manual transmission system is made up of the clutch, gearbox and driveshafts. The torque is transmitted from the engine to the road wheels via the clutch and gearbox. The normal form of clutch is referred to as a friction clutch.

The clutch

This temporarily disconnects/connects the drive between the engine and gearbox. It enables the drive to be taken up gradually.

The three main components of a clutch are the drive plate, sometimes referred to as the clutch plate or friction plate, plus the pressure plate and release bearing. The drive plate is clamped between the pressure plate and the engine flywheel by spring pressure.

The engine creates the turning motion or torque, which is transmitted from the crankshaft to the flywheel. The driveshaft, attached to the friction plate, transmits the torque to the gearbox. Depressing the clutch pedal operates the release bearing to relieve the spring clamping pressure and free the drive plate.

The life of a clutch can be prolonged by careful use and avoidance of slipping or riding the clutch. Replacement should be done before the drive plate becomes too worn, as further use could cause the flywheel to become scored.

The gearbox

The purpose of the gearbox is to

- multiply the torque (driving force) being transmitted by the engine
- provide a means of reversing the vehicle
- provide a permanent position for neutral.

The gears contained in the gearbox allow the driver to vary the speed of the road wheels corresponding to any particular engine speed. This also results in varying the tractive effort, which is applied through the tyre to the road, to overcome the resistance to the movement of the vehicle during moving off from rest, accelerating and hill climbing.

There is widespread use of semi-automatic and automatic gearbox systems to assist the driver and improve vehicle performance. In many systems there is no need for a normal clutch pedal and vehicle movement from rest is achieved in response to movement of the accelerator pedal. Gear changing may be controlled by the driver (semi-automatic) or be controlled hydraulically or, increasingly, by the use of electronic systems, to change gear according to the requirements of the vehicle use situation.

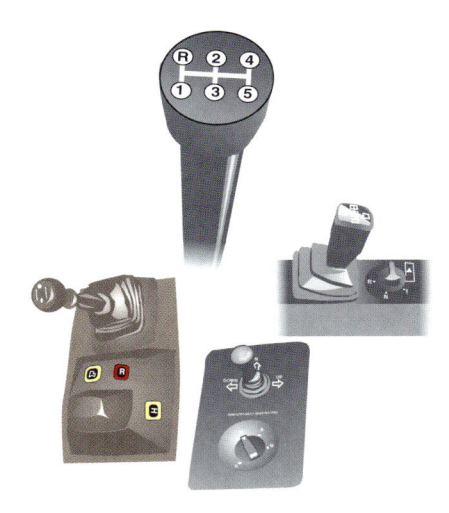

PASSENGER CARRYING VEHICLES

Electrical system

Much progress has been made regarding the systems within vehicles so that most mechanical units are now controlled by electricity. The wiring requirements are so extensive in some vehicles that a system called multiplexing is used. This system is computer controlled, it uses a cable carrying electronic messages to switch equipment on or off. A power bus cable carries the main electric current to operate the equipment.

PCV and LGV vehicles commonly use 24 volt lead/acid batteries to provide the power to start the vehicle. Once the engine is running the alternator takes over and provides the electrical power needed whilst also recharging the battery. The alternator generates electrical current and is usually directly driven by the engine via a belt. A controlled current is directed to the battery, which enables it to remain charged and provide current for other electrical systems of the vehicle, such as the lighting system.

Fuses of varying ratings, dependent on the power consumption, protect the different circuits within the vehicle. They prevent excess current from overloading the system, which may cause electrical fires. It is advisable to carry spare fuses, but make sure that you use the correct rating and find out why the fuse 'blew' before replacing it.

Care should be taken when checking batteries as explosive gases build up and the dilute sulphuric acid used as an electrolyte will burn skin. Always follow manufacturers' recommendations when dealing with batteries. There is a freely available leaflet from HSE warning of the dangers of charging batteries which gives help and advice on the correct way to deal with batteries (see page 237 for contact details).

Tyres

All tyres on your vehicle and any trailer must be in good condition. They need to be checked weekly for damage or wear and to ensure that they are at the correct pressure. Follow manufacturers' recommendations for the correct pressure required. Neglecting tyre pressures is a major cause of tyre failure: check your tyre pressures when the tyres are cold, that is, before the vehicle is used. Ensure that all tyres are suitable for the loads being carried. Passenger carrying vehicle tyres have codes indicating the maximum load and speed capability. These must be suitable for the use conditions of the vehicle.

The life of a tyre will depend upon the load, inflation pressure and the speed at which the vehicle is driven. Under-inflated tyres will increase wear of the outer edge of the tread area of the tyre. Over-inflated tyres will distort the tread and increase wear in the centre of the tread area of the tyre.

Two main types of tyre construction or structure are in common use: diagonal or cross ply and radial ply. Cross ply tyres have a large number of rubber-covered textile cords which are alternately angled across the tyre from the bead, that is the part which sits on the wheel rim on one side to the bead on the other. This results in a rather stiff side wall, which supports the tread against steering and other side forces.

Radial ply tyres have similar textile cords but arranged radially across the tyre almost at right angles to the width of the tread. The tyre walls are quite supple and a rubber covered steel mesh belt, which runs around the tyre underneath the tread rubber, braces the tread area. The belt keeps the tread in flat contact with the road to improve traction and grip.

PASSENGER CARRYING VEHICLES

Whilst both types are in use, the radial ply tyre is favoured as it gives increased tyre life, improved safety and is more widely available in tubeless form.

For safety reasons there are very strict regulations concerning the mixing of radial and cross ply tyres on different axles. The regulations are complicated for multi-axled vehicles but for a simple single front and rear axle vehicle, radial ply tyres must not be fitted to the front axle if cross ply tyres are fitted to the rear. Tyre structures must never be mixed on the same axle.

Some PCVs are designed to have different sized wheels on the front and the rear, but sizes should never be mixed on the same axle.

Keeping tyres correctly inflated will help prevent failure and also improve fuel consumption: using radial ply tyres can improve consumption by 5 to 10%.

The tread depth of tyres used on PCVs with more than 8 passenger seats must be at least 1 mm across three-quarters of the breadth of the tread, and in a continuous band around the entire circumference.

Check wheels and tyres for balance to avoid uneven wear. When a wheel and tyre rotate they are subject to centrifugal forces. If the mass of the wheel and tyre is dispersed uniformly then the wheel is balanced. Clip-on balance weights are used to rectify any imbalance.

Commercial vehicles with tubeless tyres use metal valve stems fitted to the wheel rim. Either an O-ring or a flat-flanged rubber washer makes the sealing airtight. Vehicles fitted with tube tyres have an adaptor, which is moulded to a rubber patch and vulcanised to the inner tube. The valve-stem casing is then screwed on to the tube adaptor.

Changing a tyre

Great care must be taken when changing the tyre of a large vehicle and it is often better to call out a professional tyre fitter. If you are forced to change a tyre you should

- select a firm, flat surface
- check that the parking brake is applied
- ensure the passengers or other personnel are clear of the area you are working in
- check the wheel is not damaged and that another tyre can be fitted to it
- deflate the tyre before attempting to remove the wheel
- not loosen or unscrew the clamping nuts if they are divided wheel rims
- take care not to damage the flanges and locking rings when taking the tyre off.

Fitting a new tyre

Having checked the condition of the wheel before replacing with a correctly sized tyre, you should

- renew the complete valve whenever a tubeless tyre is being replaced
- fit the wheel to the tyre whilst the wheel is laying flat on the ground. This will enable the tyre to fit the rim and obtain a good airtight seal
- inflate commercial tyres in a cage or similar safety cell
- inflate to 1 bar level with the valve core removed
- insert a valve core
- inflate to manufacturer's recommendation

PASSENGER CARRYING VEHICLES

- fully tighten wheel nuts, to the torque recommended by the vehicle manufacturer, using a calibrated torque wrench. Tighten the wheel fixings gradually and alternately diagonally across the wheel. Recheck the torque after about 30 minutes if the vehicle remains stationary or after 40 to 80 km (25 to 50 miles), if used.

Power tools are not recommended for tightening wheel fixings. It is recommended that pressure gauges are checked frequently for accuracy.

When leaving building sites or other areas with loose debris check between the tyres for bricks or other large objects that could damage your tyres or following traffic, should they fall out.

Coupling system

The coupling system is a device used to connect the PCV to a trailer. It permits articulation between the units. Guidance on the correct way to uncouple or recouple a unit can be found on page 223.

Maintenance

The coupling must be maintained properly to ensure safety. You should refer to your vehicle/trailer manual for maintenance intervals and lubricants.

Braking system

There are three braking systems fitted to PCVs.

The service brake
- The principal braking system used.
- Operated by the foot control.
- Used to control the speed of the vehicle and to bring it to a halt safely.
- May incorporate an anti-lock braking system (ABS).

The secondary brake
- May be combined with the foot-brake control or the parking brake.
- For use in the event of a failure of the service brake.
- Normally operates on fewer wheels than the service brake and therefore has a reduced level of performance.

The parking brake
- Usually a hand control. It must be a mechanical device, which may be applied or released with power assistance. Most vehicles have spring brake chambers acting on the rear axle. These cannot be released until there is sufficient air pressure (normally exceeding 60 psi) in the appropriate reservoir.
- May also be the secondary brake but should normally only be used when the vehicle is stationary.
- Must always be set when the vehicle is left unattended. (It's an offence to leave any vehicle without applying the parking brake.)

The minimum legal braking performance permitted for each system is

- service brake - 50% efficiency
- secondary brake - 25% efficiency
- parking brake - 16% efficiency.

Anti-lock braking systems (ABS)

Some vehicles are fitted with anti-lock braking systems (ABS). Wheel-speed sensors in these systems detect the moment during braking when a wheel is about to lock. Just before this happens the system reduces the braking effort and then rapidly re-applies it. This action may happen many times a second to maintain brake performance.

Preventing the wheels from locking means that the vehicle's steering and stability is also maintained, leading to safer stopping.

PASSENGER CARRYING VEHICLES

But remember, an ABS is only a driver aid. It doesn't remove the need for good driving practices, such as anticipating events and assessing road and weather conditions.

Anti-lock braking systems are commonly used on large PCVs and are required by law on some. It's important to ensure that an ABS is functioning before setting off on a journey. Driving with a defective ABS may constitute an offence.

The satisfactory operation of the ABS can be checked from a warning signal on the dashboard. The way the warning lamp operates varies between manufacturers, but with all types the light comes on with the ignition. It should go out no later than when the vehicle has reached a road speed of about 10 kph (6 mph).

Endurance braking systems

Buses and coaches are also frequently equipped with endurance braking systems (commonly called retarders). These systems provide a way of controlling the vehicle's speed without using the wheel-mounted brakes.

Retarders operate by applying resistance, via the transmission, to the rotation of the vehicle's driven wheels. This may be achieved by

- increased engine braking
- exhaust braking
- transmission-mounted electromagnetic or hydraulic devices.

Endurance braking systems can be particularly useful on the descent of long hills, when the vehicle's speed can be stabilised without using the service brake. Braking generates heat, and at high temperatures braking performance can be reduced. Proper use of endurance braking systems can prevent this from happening.

Using these braking systems will significantly reduce brake lining wear during intensive stop–start urban operation. However care must be taken to check the depletion of the air pressure in the service reservoirs due to the frequent application and release of the service brake. On long descents, the air volume usage often exceeds the replenishment rate of the compressor. This causes the service reservoir air pressure to drop below the normal maximum at which the service brake may operate.

The system may be operated with the service brake (integrated) or by using a separate hand control (independent). Retarders normally have several stages of effectiveness, depending on the braking requirement. With independent systems the driver has to select the level of performance required.

When operating independent retarders while driving on slippery roads, care must be exercised if drive-wheel locking is to be avoided. Some retarders are under the management of the ABS system to help avoid this problem.

Safety

Air brake systems are fitted with warning devices that are activated when air pressure drops below a pre-determined level in one or more air reservoirs. In some circumstances there may be sufficient pressure to release the parking brake even though the warning is showing. Under these circumstances the service brake may be ineffective. Therefore, you should never release the parking brake when the brake pressure warning device is operating.

On some vehicles a special brake may be automatically applied when the vehicle is brought to a stop. This is designed to prevent the vehicle moving until the accelerator is used to move off. This isn't a parking brake, however, so you shouldn't leave your seat until the parking brake has been applied.

PASSENGER CARRYING VEHICLES

Inspection and maintenance

You aren't expected to be a mechanic; however, there are braking system checks that **are** your responsibility.

Air reservoirs

Air braking systems draw their air from the atmosphere, which contains moisture. This moisture condenses in the air reservoirs and can be transmitted around a vehicle's braking system. In cold weather this can lead to ice forming in valves and pipes and may result in air pressure loss and/or system failure. Some air systems have automatic drain valves to remove this moisture, while others require daily manual draining. You should establish whether your vehicle's system reservoirs require manual draining and, if so, whose responsibility it is to make sure it's done.

Controls

Before each journey make sure that all warning systems are working. ABS warning signals will operate as soon as the ignition is switched on. Brake pressure warning devices may operate when the ignition is turned on (before starting the engine) or may be activated by using a special 'check' switch. **Never** start a journey with a defective warning device or when a warning is showing. If the warning operates when you're travelling, stop as soon as you can do so safely and seek expert assistance. Driving with a warning device operating may be very dangerous and is an offence.

Air brake system

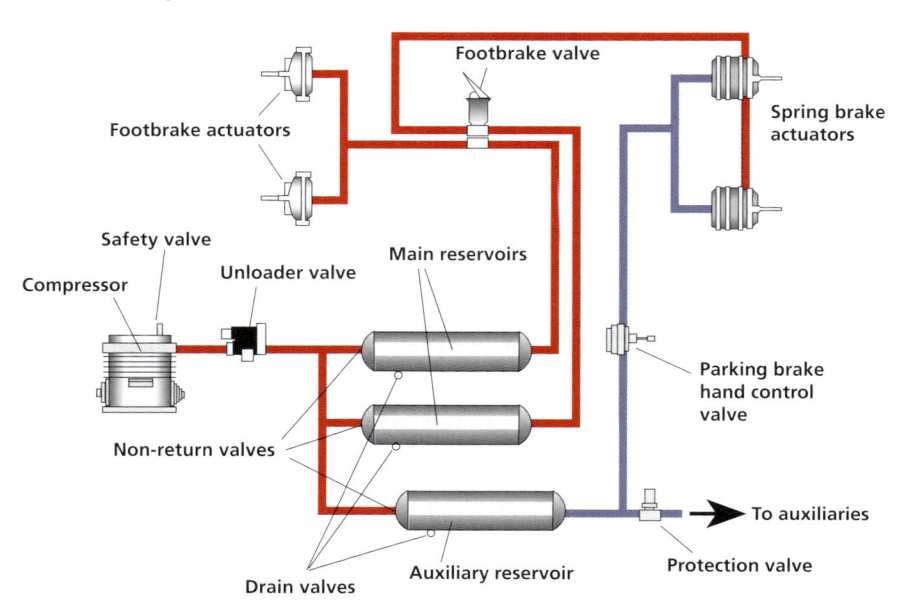

PASSENGER CARRYING VEHICLES

Other auxiliary systems
Auxiliary air systems

Modern PCVs may be equipped with air-operated accelerators, clutches, gear-change mechanisms, wipers, doors, suspension, ramps, lifts or 'kneeling' devices.

Drivers should familiarise themselves with the function and effect of these systems and be aware of any 'interlinks' that may be fitted. For example, air-operated accelerators may be disabled when the passenger doors are open.

Power-assisted steering (PAS)

Older and smaller vehicles often rely on the driver's own effort when turning the steering wheel to steer the vehicle's front wheels. So that this effort is reasonable, a gearing system is used. The driver may need to turn the steering wheel several times to reach 'full lock' (the tightest turn the vehicle can make). With historic buses it's necessary to drive more slowly round corners in order to give yourself enough time to turn the wheel.

To reduce the effort required and the amount that the driver has to turn the steering wheel, many modern minibuses, buses and coaches are fitted with a power-assisted steering system (PAS). The power assistance is often incorporated within the steering box.

PAS reduces the driver's efforts when turning. However, it only operates when the engine is running. If a fault develops you can retain control of the steering, but much greater effort is needed to turn the steering wheel. Movement at the steering wheel may also be felt as a series of jerks.

Don't attempt to drive a vehicle fitted with PAS

- without the engine running – that is, 'coasting'
- if the system is faulty.

If a fault develops whilst travelling, stop as soon as you can safely do so and seek expert assistance.

Part Three

Limits and regulations

The topics covered

- Basic knowledge
- Environmental issues
- Drivers' hours
- Other regulations
- Anti-theft measures

LIMITS AND REGULATIONS

Basic knowledge

The passenger transport industry is subject to a wide range of regulations and requirements relating to

- drivers
- operators
- companies
- vehicles
- passengers
- workshops.

The first thing you'll need to know about is your vehicle. The various aspects to consider are its

- weight (restrictions)
- height (clearances, etc.)
- width (restrictions)
- length (lay-bys, corners)
- ground clearance (for humpback bridges, grass verges, kerbs, etc.)

You'll also need to know the various speed limits that apply to your vehicle and the speeds at which it will normally travel and cruise

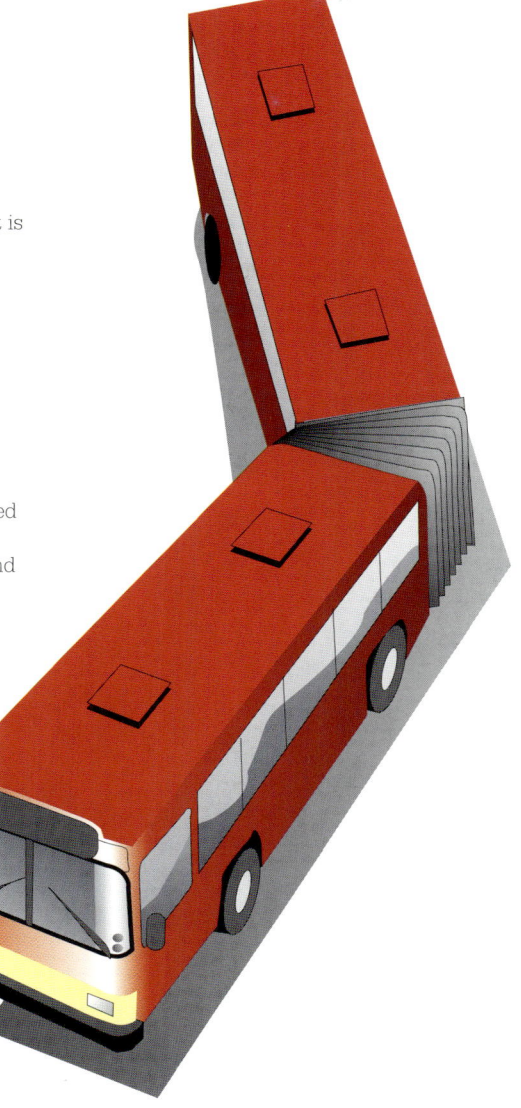

57

LIMITS AND REGULATIONS

Weight

Weight limits are imposed on roads and bridges for two reasons

- the structure may not be capable of carrying greater loads
- to divert larger vehicles to more suitable routes.

Sometimes buses and coaches are exempt from the notified limits by means of a plate beneath the weight limit sign. This normally refers to PCVs in service or requiring to use that particular road for access. If you can use another route, do so. Remember, try to be considerate towards local people and the environment.

You should be aware of, and understand, the limits relating to any vehicle you drive. Certainly you should make sure that you know what your vehicle weighs.

In many cases, weight limits apply to the maximum gross weight (MGW). To arrive at this figure add about 1 tonne per 15 passengers to the unladen weight shown on your vehicle, plus an allowance for any luggage you may be carrying. For example

75-seat double-deck coach	12.0 tonnes
75 passengers	5.0 tonnes
75 cases	1.5 tonnes
500 litres fuel	0.5 tonnes
Total weight	**19.0 tonnes**

The weight difference between a laden and unladen coach may be as much as 7 tonnes.

The loading and distribution of large amounts of luggage can affect axle weights and stability.

Definitions of terms to do with weight limits can be found in the Glossary at the back of this book.

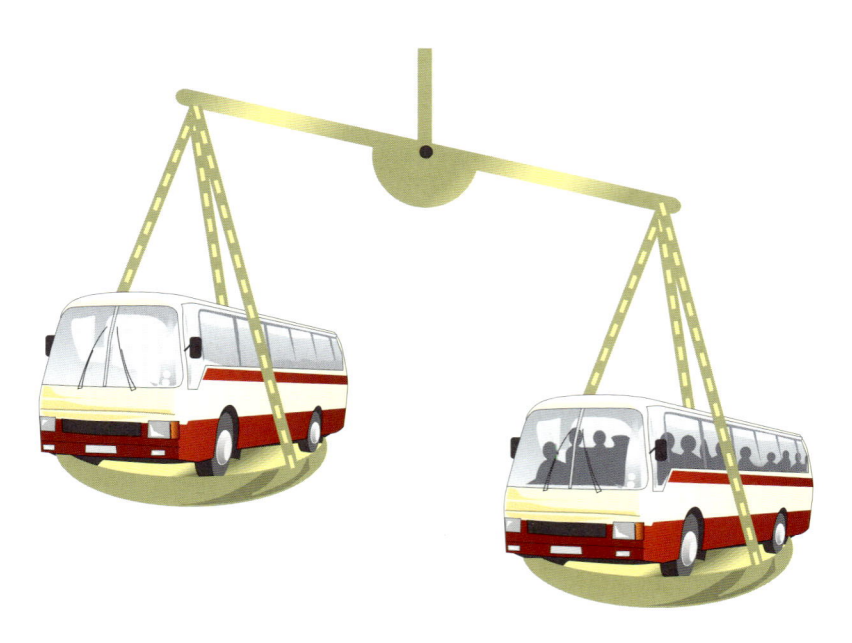

LIMITS AND REGULATIONS

Height

You aren't allowed to drive a vehicle that has an overall travelling height of more than 3.0 metres (9 feet 10 inches) unless the overall travelling height (including any trailer) is conspicuously marked

- in feet and inches, or in feet and inches and in metres so that there's no more than 50 mm difference between the height specified in feet and inches and the height specified in metres
- in figures at least 40 mm high, which can be read by the driver when in the driving position.

A driver may drive a vehicle higher than 3.0 metres (9 feet 10 inches) where the route is known, for example a local bus service, or if there is easy access to a document on the vehicle describing the safe route.

In addition, you should ensure that

- any height indicated isn't less than the overall travelling height of the vehicle
- this is the only indication of the overall travelling height.

Overhead clearances

Drivers of any vehicle exceeding 3.0 metres (9 feet 10 inches) in height should exercise care when entering

- loading bays
- bus and coach stations
- depots
- refuelling areas
- service station forecourts
- any premises that have overhanging canopies

or when driving under or negotiating

- bridges
- overhead cables
- overhead pipelines
- overhead walkways
- road tunnels.

The normal maximum permitted overall travelling height of any PCV with fixed bodywork is 4.57 metres (15 feet). Many countries in the EC don't allow PCVs in excess of 4.0 metres (13 feet) without an exceptional vehicle permit being applied for, and issued, in advance.

Be aware of overhanging tree branches, particularly on roads rarely used by high vehicles, in case upper-deck windows are broken. Trees on regularly used routes are generally kept trimmed. If in doubt, slow right down and, if necessary, stop, get out and check.

Don't take chances.

In addition, most roads have a slope (camber) to help with drainage, and this can sometimes cause problems. For example, on roads with a severe camber, the top of a double-deck bus can lean up to 250 mm (around 10 inches) further over than the wheels. This situation could be made worse when pulling up at bus stops, if the nearside wheels drop into the gutter. Lamp-posts, traffic signs, shop awnings, bus shelters, etc. are within this 'danger zone', so watch out for these hazards.

LIMITS AND REGULATIONS

Bridges

Every year there are about 800 accidents where vehicles hit railway or motorway bridges, some involving buses and coaches. This means that, on average, there are **more than two incidents every day of the year.**

Collisions involving buses can kill or injure passengers, not to mention weakening the bridge. When a railway bridge is involved, such a collision frequently disrupts rail traffic and could lead to a major disaster. There are also the additional costs involved in making the bridge safe, re-aligning railway tracks, etc., apart from the general disruption to road and rail traffic.

The headroom under bridges in the UK is at least 5 metres (16 feet 6 inches), unless otherwise indicated. Where the overhead clearance is arched this is normally **only** between the limits marked.

You **must** know the height of your vehicle: don't guess. If in doubt, measure it or look at the information shown in the cab.

Always be on the lookout for height restrictions shown on

- traffic signs
- road markings
- warning lights.

Stay alert to the dangers.

60

LIMITS AND REGULATIONS

If your vehicle collides with any bridge STOP. Your first responsibility is to your passengers, so check that there are no injuries. If there are, take appropriate action (see the advice on first aid in Part Four).

You must report the collision to the police. If a railway bridge is involved, report it to the railway authority as well, by calling 0845 711 4141.

Do this immediately, to avoid a possible serious accident or loss of life.

Give information about

- the location
- the damage
- any bridge reference number (often found on a plate bolted to the bridge or wall).

You must inform the police within 24 hours, if you don't do so at the time of the incident. Failure to notify the police is an offence.

If you are not sure of the safe height of a railway bridge, stop and call Railtrack. To avoid problems

- plan your route carefully
- slow down when approaching bridges
- know the height of your vehicle
- keep to the centre of arched bridges
- wait for a safe gap to proceed if there's oncoming traffic.

Height guide

Metres	Feet/Inches
5.0	16 6
4.8	16
4.5	15
4.2	14
3.9	13
3.6	12
3.3	11
3.0	10
2.7	9

LIMITS AND REGULATIONS

Width

You must always be aware of the road space your vehicle occupies. This is particularly important where road width is restricted because of parked or oncoming vehicles, or in narrow roads.

Many local authorities now use 'traffic calming' measures, which often include road width restrictions. Watch out for these. If you know of roads with such restrictions, try to avoid them, unless you're following a scheduled service route, of course.

The majority of buses and coaches in the UK are 2.5 metres wide (8 feet 3 inches) but the legal maximum width is 2.55 metres (8 feet 5 inches). Mirrors and exterior trim can also add to a vehicle's width.

Where space is limited, take particular care when meeting other large vehicles. If necessary, stop first and, only if you're certain there's enough space, manoeuvre past slowly. Keep a lookout all round and especially watch out for mirrors hitting each other or lamp-posts, etc. A broken mirror means that your vehicle is unroadworthy and, therefore, illegal. It could also cause injury to you or others.

LIMITS AND REGULATIONS

Length

You need to know the length of your vehicle, as well as its width, so that you can judge the space you need on the road. You'll also need to know these dimensions to comply with regulations that affect your vehicle.

Other than 'traffic calmed' zones, places where there are restrictions on vehicle length are comparatively rare. Examples are

- road tunnels
- level crossings
- ferries.

The usual maximum length for a bus or coach is 12 metres (39 feet 4 inches). Articulated buses may be up to 18 metres long (59 feet).

Drivers of long vehicles must be careful when

- turning left or right
- negotiating roundabouts or mini-roundabouts
- emerging from premises or exits
- overtaking
- parking, especially in lay-bys
- driving on narrow roads where there are passing places
- negotiating level crossings.

Be aware of the amount of space you need to turn (the 'turning circle') and the way that your vehicle overhangs kerbs and verges (the 'swept area').

You must be particularly aware of the risk of grounding, for example, on a hump bridge, and you should look out for appropriate traffic warning signs.

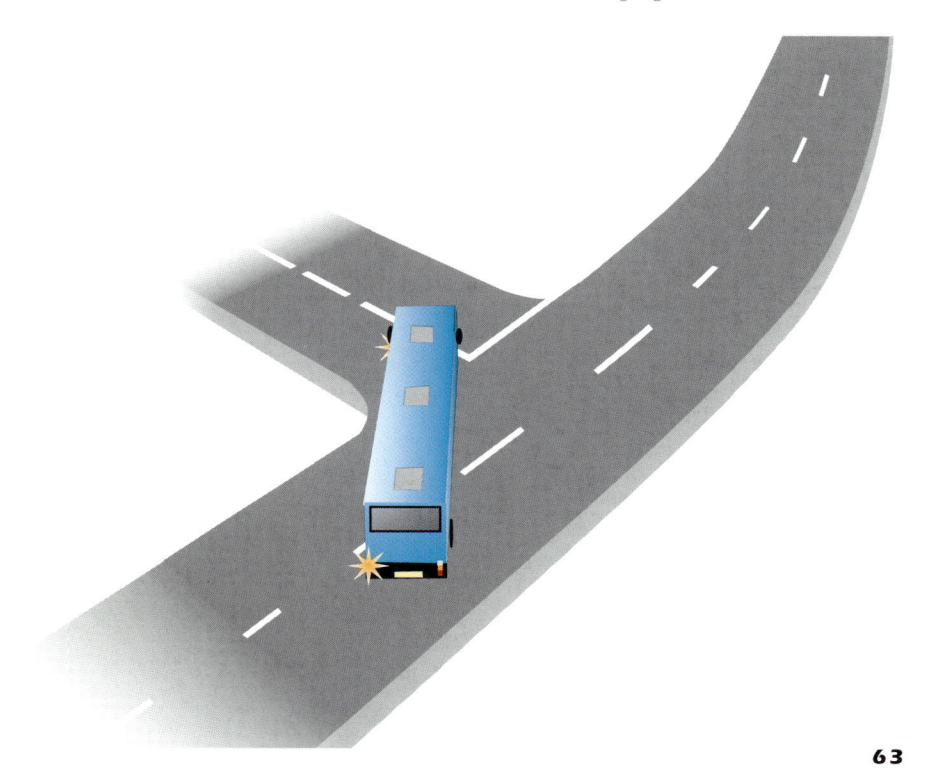

LIMITS AND REGULATIONS

Environmental issues

Vehicle designers, bus operators and maintenance staff all have a part in helping to reduce the effects that vehicles have on the environment. You can also help. You should be aware of the effects your vehicle, and the way in which it's driven and operated, can have on the environment around you.

The bus and coach industry has a major role to play in limiting the total number of vehicles on our roads. One double-decker bus can carry the occupants of 20 cars. Therefore, only one engine could be running instead of 20. However, a badly maintained or poorly driven bus can cause unnecessary pollution, perhaps as much as several cars.

What YOU can do to help:

- Follow manufacturer's recommendation for servicing.
- If you do your own maintenance, make sure you take your old oil, batteries and used tyres to a garage or local authority site for recycling or safe disposal. It is illegal and harmful to pour oil down a drain.
- Make regular checks of your vehicle and ensure that any defects are reported and sorted out.
- Check excessive exhaust smoke (the public is encouraged to report vehicles emitting excessive fumes).
- Check uneven running, which may be caused by fuel pump or injector faults.
- Check brake faults, which can cause drag.
- Have correct tyre pressures.
- Check suspension system faults, which may result in road damage.

Always drive with fuel economy in mind. Operators keep careful checks on vehicle running costs, and fuel economy is a key factor for profitability as well as reducing waste.

You should

- plan routes to avoid congestion
- anticipate well ahead
- avoid the need to 'make up time'
- avoid over-revving (if a rev counter is fitted try to keep in the green band as much as possible)
- drive smoothly. This can reduce fuel consumption by 15 per cent. Avoid rapid acceleration as this leads to greater fuel consumption, wear and tear
- avoid using the air-conditioning continuously as this increases fuel consumption by about 15 per cent
- brake in good time (all braking wastes energy in the form of heat)
- make good use of regenerative retarders where fitted. This is a braking system which allows the use of the vehicle's drive motor, or motors, to convert the vehicle's kinetic energy into electrical energy during deceleration
- switch off your engine when stationary for some time, especially where noise and exhaust fumes cause annoyance
- allow air pressure to build up with the engine on 'tick over' rather than 'revving up'.

LIMITS AND REGULATIONS

Select for economy and low emissions

- Vehicles with automatic transmission use about 10 per cent more fuel than those with manual transmission.
- Consider using ultra-low sulphur fuel, such as city diesel, as it reduces harmful emissions.
- When replacing tyres, consider buying energy-saving types. These have reduced rolling resistance, and they increase fuel efficiency and also improve grip on the road.

Further information and publications can be found on TransportEnergy's website, www.transportenergy.org.uk.

Traffic management

Continuous research has resulted in new methods of helping the environment by easing traffic flow.

Traffic flow

Strict parking rules in major cities and towns help the traffic flow. Red Routes in London have improved the traffic flow considerably.

Speed reductions

Traffic calming measures, including road humps and chicanes, help to keep vehicle speeds low in sensitive areas. Do not speed up between the road humps. There is an increasing number of areas with a 20 mph speed limit in force.

For general enquiries about traffic calming measures contact the Department for Transport, tel. 020 7944 2594.

LIMITS AND REGULATIONS

Road-friendly suspension

Bumping your vehicle over kerbs, verges and pavements damages them and can also affect underground services. Repairs can be costly. Many PCVs are fitted with air suspension to reduce wear on road surfaces.

Damage to your vehicle's tyres, which may not be immediately obvious, can also be the result of poor driving and bad suspension. Subsequent tyre failure may have serious consequences, possibly to another driver and his or her passengers. Make sure that you drive responsibly and with due care, even if your vehicle is fitted with road-friendly suspension.

Fuels

Take care to avoid spillages when you refuel your vehicle. Diesel fuel is slippery and can be very dangerous if stepped on (especially in garage areas). On the road it can create a serious risk to other road users, especially motorcyclists. It's a legal requirement that you check all filler caps are properly closed and secure before driving off.

Exhaust emissions

Fuel combustion produces carbon dioxide, a major greenhouse gas, and transport accounts for about one-fifth of the carbon dioxide we produce in this country.

MOT tests now include a strict exhaust emission test to ensure that all vehicles are operating efficiently and causing less air pollution.

Diesel engines

These engines are more fuel-efficient than petrol engines. Although they produce higher levels of some pollutants (nitrogen oxides and particulates), they produce less carbon dioxide (a global warming gas). They also emit less carbon monoxide and hydrocarbons.

Alternative fuels

To improve exhaust emissions even further, ultra-low sulphur diesel or 'city diesel' fuels can be used. These have been formulated so that the sulphur content is very low. Sulphur is the main cause of particulates in exhaust emissions and it also produces acid gases. The lower the content of sulphur in fuel, the less damage to the environment.

Electricity
Trials have been taking place with electric vehicles for a number of years, but it is only recently that advances have been made in overcoming the problems of battery size and capacity.

Fuel cells
These operate like rechargeable batteries and produce little or no pollutants, but have greater range and improved performance than most battery electric vehicles.

Hybrid vehicles
These offer the advantages of electricity without the need for large batteries. The combination of an electric motor and battery with an internal combustion engine gives increased fuel efficiency and greatly reduced emissions.

Hydrogen
This is another possible fuel source for road vehicles that is being studied. However, technical problems include storage of this highly inflammable gas.

LIMITS AND REGULATIONS

Liquid Petroleum Gas (LPG)
This consists mainly of methane, produced during petrol refining. Vehicles can run on LPG alone or both LPG and petrol (known as 'dual fuel'). Most types of engines can be built or converted to run on LPG. Benefits include low cost, lower emissions and reduced wear and tear to engine and exhaust systems. Disadvantages include cold start problems and valve-seat wear.

Methane
Because of the naturally occurring renewable sources of this fuel, it is also being considered as a possible alternative to diesel oil, which is a finite resource.

Solar power
Needing only daylight to function, solar vehicles are small, light, slow and silent. They produce no emissions at all; however, they are very expensive as yet and improvements are needed so they can store energy for use in the dark.

Audible warning systems

Some vehicles are fitted with systems which warn people that the vehicle is reversing, such as

- bleepers
- horns
- voice warnings.

These must not be allowed to operate on a road subject to a 30 mph speed limit between 11.30 pm and 7 am. And remember, using an audible warning device doesn't take away the need to practise good, all-round effective observation.

Also, take care when setting vehicle alarm systems. There are restrictions on the length of time that the warning may sound. Environmental Health officers are empowered to enter vehicles and disable the system if a nuisance is caused.

LIMITS AND REGULATIONS

Drivers' hours

Drivers' hours are controlled in the interest of road safety, drivers' conditions and fair competition. European regulations set maximum limits for driving times and minimum requirements for rest breaks. These are known as EC rules. Breaking these rules will result in heavy fines and you may lose your licence. It is illegal to tamper with, or alter with intent to deceive, drivers' hours records.

Tachographs

A tachograph is a device that records hours of driving, other work, breaks and rest periods. It can also record the vehicle's speed and distance travelled. The tachograph should be properly calibrated.

Tachographs are needed under EC rules for all vehicles with

- more than 16 passenger seats, except for those on regular services
- more than eight passenger seats on EC international journeys.

There are regulations that require tachographs to be

- correctly calibrated before first use
- inspected every two years
- recalibrated every six years.

Charts

Although there are exceptions, the basic requirements are that

- you must carry enough charts of an approved type. You'll need one for every 24 hour period
- your employer is responsible for supplying charts which can be used in the instrument fitted to the vehicle. They must supply a sufficient amount for the whole journey, as well as spares. This is in case any get damaged or are taken by an authorised inspecting officer
- dirty or damaged charts must not be used
- record sheets for the current week and the last day of the previous week must be carried
- record sheets must be returned to the operator within 21 days of use.

For more detailed information on drivers' hours see the DfT booklet PSV 375 entitled Drivers' Hours and Tachograph Rules for Road Passenger Vehicles in the UK and Europe. Copies of the booklet can be obtained from local Traffic Area Offices (see page 233 for contact details).

LIMITS AND REGULATIONS

Recording information

The driver must fill in all the information detailed in the centre field of each chart before use.

You must record all periods of work on the chart, ensuring the correct setting of the 'mode' switch at all times.

This switch shows when the driver is

- driving

- doing other work

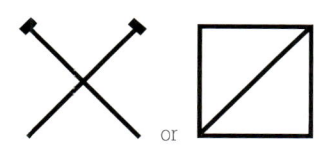

- on duty and available for work

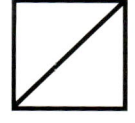

- taking a break or rest.

If you're working away from the vehicle and can't leave a chart in the tachograph – or have left the chart in but you've changed your work mode while away from the vehicle – you must make a manual entry on the reverse of the chart to that effect.

Chart inspection

Employers must make regular checks to ensure that the rules are being obeyed. The charts must be kept for at least one year after their use and submitted to enforcement authorities as required.

LIMITS AND REGULATIONS

EC drivers' hours

'Driving' means being at the controls of a vehicle for the purposes of controlling its movement, whether it's moving or stationary with the engine running.

Daily driving

A day is defined as any period of 24 hours beginning when you start other work or driving after the last daily or weekly rest period. The maximum daily hours you may drive is nine. This can be increased to 10 hours twice a week. The basic nine hours must be between

- two daily rest periods, or
- a daily rest period and a weekly rest period.

Any driving off the public roads doesn't count as driving time. In this case you should record the time as 'other work'.

You must ensure that you take a break of 45 minutes after four and a half hours of driving. This break can be replaced by two or three breaks of no less than 15 minutes during or after the driving period. The total of these shorter breaks must add up to at least 45 minutes in the four and a half hours of driving. During any break you must not drive or undertake any other work.

Daily rest periods

A rest period is an uninterrupted period time of at least one hour during which you may freely dispose of your time.

You must have a minimum of 11 consecutive hours' daily rest. This can be reduced to nine hours, but not more often than three days a week. In this case you must compensate for the reduction by taking an equivalent rest before the end of the following week.

Daily rest can alternatively be taken as 12 hours in two or three periods. In this case each rest period must be at least one hour and the last period must be at least eight consecutive hours.

If you're taking your rest period on a ferry or train the daily rest period may be interrupted for up to one hour and includes any customs formalities, but only once. If it is, two hours must be added to the rest time. One part must be taken on land, either before or after the journey. The other part can be taken on board a boat or train. In this case you must have access to a bunk or couchette for both rest periods.

Weekly driving

A week is defined as a period between 0.00 hours on Monday and 24.00 on the following Sunday.

There is no weekly driving limit, but a weekly rest period must be taken after no more than six daily driving periods (or the twelfth day on non-regular national or international passenger services). You can drive up to 56 hours between weekly rest periods, but must not exceed the limit of 90 hours in any one fortnight.

Weekly rest periods

When taking the weekly rest period, a daily rest period must normally be extended so that you get at least 45 consecutive hours of rest. You can reduce this to a minimum of 36 hours if you take the rest either where the vehicle is normally based or where you're based. If it's taken elsewhere it can be reduced to a minimum of 24 consecutive hours.

If you take reduced rest you must make up for it by taking an equal period of rest added to a weekly or daily rest period. This must be taken in one continuous period before the end of the third week following the week in question.

LIMITS AND REGULATIONS

A weekly rest period that begins in one week and continues into the following week may be added to either of these weeks.

Catching up on reduced rest

If you've reduced your daily and/or weekly rest periods the compensatory rest must be added to another rest of at least eight hours. You can request to take this at either your base or where your vehicle is based. Rest taken as compensation for the reduction of a weekly rest period must be taken in one continuous block. Rest taken as compensation for the reduction of a daily rest period can be made up of any combination of breaks of at least one hour.

Two or more drivers

During each period of 30 hours, each driver must have a rest period of not less than eight consecutive hours. There must always be two or more drivers travelling with the vehicle for this rule to apply. A driver may take a break while another driver is driving, but not a daily rest period.

In the interest of road safety all rules regarding drivers' hours should always be followed. However, there might be an emergency situation where you have to depart from the drivers' rules to ensure the safety of people, the load or the vehicle. In these unusual situations you should note the reasons on the back of the tachograph chart.

Using a ferryboat

If a driver covers part of the journey on a ferryboat or train the daily rest period may be interrupted, but only once, and if it is, two hours must be added to the total rest time. If the rest period is split, one part must be taken either before or after the journey on land. During both parts of the rest access must be available to a bunk or couchette.

Regular services

A regular service on a route of over 50 km is subject to EC rules, but a tachograph isn't needed as long as

- the employer draws up a service timetable and duty roster for crew members, records of which are kept for inspection
- the driver carries an extract from the duty roster and a copy of the service timetable.

A regular service on a route of up to 50 km is free from the EC rules but will, in most cases, be subject to the domestic drivers' hours rules.

Vehicles operating services under the 'permit' scheme may not require tachographs or be subject to the EC rules. The requirements will be explained when the permit is issued and will depend on the use made of the vehicle.

Domestic drivers' hours

The domestic rules apply to those vehicles on journeys within the UK that are specifically exempted from the EC rules. No written records are required under these rules.

71

LIMITS AND REGULATIONS

Mixed EC and domestic rules

You may find yourself working partly under EC rules and partly under UK domestic rules (sometimes even on the same day). In situations where you work 'mixed' hours you must know which set of rules to apply.

Remember the following points

- when driving under each set of rules you must comply with the requirements of the specific rules being followed
- time spent driving or on duty under one set of rules can't count as a break or rest period under the other set of rules
- driving and other duties under EC rules count towards the limits on driving and other duties under UK domestic rules
- driving and other duties under UK domestic rules (including non-driving work in another employment) count as 'attendance at work' under EC rules.

Recording 'mixed' hours

Tachographs aren't required under UK domestic rules, but you'll need to make a manual entry on your tachograph chart, showing periods of domestic driving as 'other work', when driving under EC rules.

If you're using a tachograph when driving under domestic rules the mode switch should be set on 'other work' (see page 69 for more information).

Unforeseen events and emergencies

Employers must schedule work to enable drivers to comply with the EC rules on drivers' hours. However, providing road safety is not jeopardised, and to ensure the safety of persons, vehicle or load, a driver may depart from the rules in order to reach a suitable stopping place. Reasons for doing so must be recorded on the back of the tachograph record sheet. This should not be a regular or repeated occurrence, as it would indicate work was not being correctly scheduled. Planned breaches of the driver's hours are not permitted.

LIMITS AND REGULATIONS

Other regulations

Public Service Vehicle Operator Licensing

A Public Service Vehicle Operator's Licence is required for any vehicle that carries passengers by road for payment (this is called 'Hire or Reward'). Hire or Reward is any sort of payment which gives a person a right to be carried on a vehicle regardless of whether a profit is made or not. The payment may be made by the actual person, or on their behalf, and may be a direct payment (e.g., a fare) or an indirect payment (such as membership subscription to a club, payment for a room in a hotel or school fees).

For further information on the requirement to hold a PSV Operator's Licence, contact your local Traffic Area Office. For contact details, see pages 233-234.

Northern Ireland operations

In Northern Ireland anyone who carries passengers by road for reward must hold a Road Service Licence, which is granted by the Road Transport Licensing Division of Driver and Vehicle Licensing Northern Ireland.

Know the regulations

In Britain during 2001, almost 10,000 passengers and drivers became casualties in collisions involving buses and coaches. In the same year, ten passengers and four drivers died and there was an average of 31 incidents a day involving PCVs.*

A small, but significant, proportion of such incidents involve a PCV driver breaking the law in some way. Drivers are also prosecuted for offences where no accident has occurred. Don't let this happen to you.

You must comply with regulations that affect

- your health and conduct
- your vehicle
- your driving
- your passengers
- health and safety issues.

It is essential that you know and keep up to date with the regulations and the latest official advice.

*Note
These figures are taken from Road Accidents Great Britain: 2001, The Casualty Report, which is published by the Stationery Office for the Department for Transport.

LIMITS AND REGULATIONS

Your health and conduct

Health

Even apparently simple illnesses can affect your reactions. You should be on your guard against the effects of

- a common cold
- flu symptoms
- hay fever
- tiredness.

Falling asleep

Incidents where vehicles have

- left the road
- collided with broken-down vehicles, police patrol officers or other people on the hard shoulders of motorways

have now been attributed to the problem of drivers falling asleep at the wheel.

Be on your guard against boredom on comparatively empty roads or motorways, especially at night. Always

- take planned rest breaks
- keep a plentiful supply of fresh air circulating around the driving area
- avoid allowing the driving area to become too warm
- avoid driving if you aren't 100 per cent fit
- avoid driving after a heavy meal.

As a professional driver you must make sure you are always fit and able to concentrate for the whole of your shift.

A relief driver should be used at any time if you feel unwell.

Modern vehicles with air suspension, power steering and automatic transmission are less demanding to drive, but road and traffic conditions require full concentration at all times.

LIMITS AND REGULATIONS

Alcohol

It's an offence to drive with

- a breath alcohol level in excess of 35 µg per 100 ml
- a blood alcohol level in excess of 80 mg per 100 ml.

Alcohol may remain in the body for around 24 hours. So, you could still fail a breath test the morning after drinking. The only safe limit, ever, is a zero limit.

Don't drink if you're going to drive.

If you're convicted of a drink–driving offence while driving a car, any subsequent driving ban will mean that you also lose your PCV entitlement. That could mean losing your job. It's not worth the risk.

Drugs

Some operators, concerned about drug abuse amongst staff, have introduced random drug-testing for their drivers. Drivers who fail such tests face instant dismissal.

You must not take any of the following drugs, classified as 'banned substances', whilst driving

- amphetamines (e.g., 'diet pills')
- barbiturates (sleeping pills)
- benzodiazepine (tranquillizers)
- cannabis
- cocaine
- heroin
- methaqualone (sleeping pills)
- methylamphetamines (MDMA)
- morphine/codeine
- phencyclidine ('Angel Dust')
- propoxyphane.

Unlike alcohol, the effects of which last for about 24 hours, many drugs remain in the body for up to 72 hours. Even everyday cold or flu remedies can cause drowsiness. Read the label of any medicines carefully.

If in doubt, consult either your doctor or pharmacist. If still in doubt, **don't drive.**

LIMITS AND REGULATIONS

Your vehicle

The law relating to vehicles is extensive. Manufacturers, operators and drivers all must obey specific regulations.

The manufacturer is responsible for ensuring that the vehicle is built to comply with the Construction and Use Regulations.

The operator is responsible for making sure that a vehicle

- continues to comply with those regulations
- meets all current requirements and new regulations as they're introduced
- is tested as required
- displays all required markings, discs and certificates
- is in a serviceable condition, including equipment, fittings and fixtures.

In addition, the operator must operate a system whereby drivers of the vehicle can report defects and have them solved effectively. The operator shouldn't cause or permit a vehicle to be operated in any way other than the law allows.

The driver has a legal responsibility for

- taking all reasonable precautions to ensure that legal requirements are met before driving any vehicle

- checking that the vehicle is fully roadworthy and free from significant defects before driving it
- ensuring that any equipment, fittings or fixtures required are present and serviceable
- not driving the vehicle if any fault develops that would make it illegal to be driven
- ensuring that all actions taken whilst in charge of the vehicle are lawful.

You should consider whether it would be illegal to drive the vehicle if anything that should by law be fitted to or carried on the vehicle isn't in place or in a serviceable condition.

Similarly, if something is fitted to the vehicle which isn't required by law but is

- unserviceable
- in a dangerous condition
- not fitted so as to comply with the regulations

you should consider its legal status. For example, your vehicle isn't required by law to have spot or front fog lights. However, if they're fitted they must be positioned no less than 0.6 m (2 feet) from the ground.

LIMITS AND REGULATIONS

Daily walk-round checks

A daily walk-round check must be undertaken and should cover

- brakes
- lights
- tyres
- windscreen wipers and washers
- horn
- mirrors
- speedometer
- tachograph
- number plates
- reflectors and reflective plates
- exhaust system
- any coupling gear
- speed limiter
- seat belts (if fitted)

Any defects must be reported. Make sure you know the defect reporting procedure.

VOSA checks

The Vehicle and Operator Services Agency (VOSA) carries out frequent spot checks of vehicle condition. Where serious defects are found, the vehicle is prohibited from further use until the defects are rectified. Details of the prohibition are notified to the Traffic Commissioner.

Cockpit drill

Make these checks for the safety of yourself, your passengers and other road users. Every time you get into your vehicle, check that

- the driving seat is correctly adjusted, so that you can sit with a correct posture, reach all controls comfortably and take effective observations
- all interior and exterior mirrors are clean and correctly adjusted
- lenses and screens of rear-view video equipment are clean and clear
- gauges and warning systems are working correctly (never start a journey with a defective warning device or when a warning light is showing)
- the parking brake is applied
- the gear selector is in neutral (or in 'Park' if driving an automatic vehicle)
- you have sufficient fuel for your journey
- your mobile phone is switched off
- the doors are working correctly and are closed before moving off.

Before starting your journey, make sure you know and understand the

- controls: where they are and how they work
- vehicle size: its width and height, and its weight
- handling: the vehicle's characteristics
- brakes: whether ABS brakes are fitted.

LIMITS AND REGULATIONS

Speed limiters

EC legislation requires that buses and coaches over 7.5 tonnes maximum authorised mass (MAM) and capable of speeds over 100 kph (62 mph) must be fitted with a speed limiter set to 100 kph if first used on or after 1 January 1988 on

- national journeys
- international journeys.

The speed at which the limiter is set must be shown on a plate displayed in a conspicuous position in the driver's cab.

As a result of the fitting of speed limiters set to 100 kph (62 mph), buses and coaches may not use the far right-hand lane on a motorway. The restriction does not apply to A class roads with three lanes.

Your driving

When driving, it's your responsibility to follow all the relevant regulations. You must keep up to date with the road traffic rules and apply them.

This book covers the approach that you should take as a professional PCV driver. It doesn't include general driving principles – that is, everything that you should know and apply when driving **any** vehicle.

You may wish to refer to other DSA publications, such as *The Official Driving Test* and *Driving - the Essential Skills*. You should also refer to *The Highway Code* to keep up to date with revisions in traffic rules and any new road signs that may be introduced. Books by other publishers also cover general driving rules and regulations.

Keep in mind that ignorance of the law is no defence. It's reasonable to expect that you, as a professional driver, will be knowledgeable.

LIMITS AND REGULATIONS

The speed limit for buses and coaches not exceeding 12 metres in overall length is

- 70 mph on motorways
- 60 mph on dual carriageways
- 50 mph on any other road unless another, lower speed limit applies.

The speed limit for buses and coaches over 12 metres in length and for PCVs with a trailer is

- 60 mph on motorways and dual carriageways
- 50 mph on any other road unless another, lower speed limit applies.

Buses and coaches with two or more trailers are limited to

- 40 mph on motorways
- 20 mph on any other road unless another, lower speed limit applies.

Speeding offences

Many police forces and local authorities now use up-to-date technology to persuade drivers to obey speed limits, and to catch and prosecute those who don't.

Sophisticated detection equipment can 'lock on' to individual vehicles in busy traffic flows. Cameras can photograph vehicles exceeding the speed limit. At some motorway sites, roadside detection equipment displays the registration number and speed of vehicles to 'show up' the drivers concerned.

Speeding drivers who've been prosecuted find that the penalties are often linked to how much the legal speed limit was exceeded. But remember, the aim is to improve driving standards, not to increase prosecutions.

Your driving licence

You need your PCV licence in order to earn your living driving buses, coaches and minibuses. To keep it you'll want to drive to a high professional standard.

When you drive any other vehicle – your own car, for example – it's essential that your driving continues to be to the same high standard.

Your PCV licence will be at risk if you accumulate penalty points from offences committed whilst driving **any** vehicle.

Speed limits

Your vehicle will probably be fitted with a speed limiter, which will generally prevent you from exceeding motorway speed limits. However, it won't stop you exceeding lower speed limits. Observing speed limits is part of your responsibility.

LIMITS AND REGULATIONS

Red light cameras

Cameras are increasingly being installed at light-controlled junctions to record drivers who don't comply with the signals. These cameras are also intended to act as a deterrent and to improve safety in general for road users.

Any photograph produced as evidence and that shows the

- time
- date
- speed
- vehicle registration number
- length of time a red signal had been showing

will be difficult to dispute.

```
TIME:    1402
DATE:    27.02.94
SPEED:   32.7 MPH
NO. CAM: BD 37057
```

80

LIMITS AND REGULATIONS

Red Routes

On many roads in London yellow lines have been replaced with red lines. A network of priority (Red) routes for London was approved by Parliament in June 1992 as a means of addressing traffic congestion problems and widespread disregard of parking restrictions in the capital. Red Route measures currently apply to 580 miles of London's roads.

Yellow-line exemptions **don't** apply on Red Routes. During the day loading is only allowed in marked boxes. Overnight and on Sundays most controls are relaxed to allow unrestricted stopping. It's important to check signs carefully as the hours of operation for Red Routes vary from area to area.

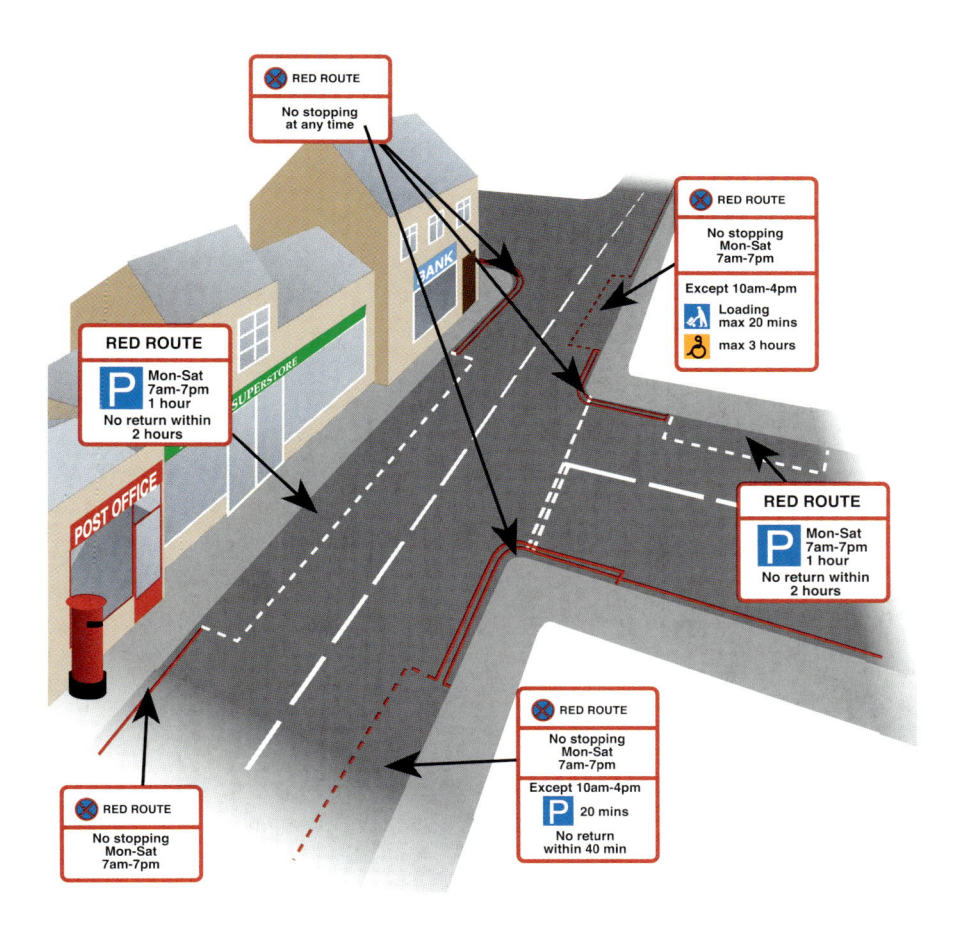

81

LIMITS AND REGULATIONS

Red Route controls are enforced by Metropolitan Police traffic wardens. There's a fixed fine for illegal stopping on a Red Route, with no discounts for early payment.

The police or traffic wardens are able to provide limited dispensations for the rare occasions when loading provisions are not adequate. These will be available from the local police station.

There are five main types of Red Route markings.

Double red lines These ban all stopping 24 hours a day, seven days a week. You're not allowed to stop for

- loading
- dropping off passengers
- visiting shops.

Single red lines These ban all stopping during the daytime, such as 7 am to 7 pm Monday to Saturday. Outside these hours unrestricted stopping is allowed.

Parking boxes These allow vehicles free short term parking and can also be used for loading. Red boxes allow parking or loading outside rush hours such as between 10 am and 4 pm for either 20 minutes or an hour during the day. White boxes allow parking or loading at **any** time but the length of stay may be restricted to 20 minutes or an hour during the day. At other times, such as 7 pm to 7 am and on Sundays, unrestricted stopping is allowed in either type of parking box.

Loading boxes These mark the areas where only loading is allowed. Red boxes allow loading outside rush hours, such as between 10 am and 4 pm, for a maximum of 20 minutes. White boxes allow loading at **any** time, but during the day the length of stay is restricted to a maximum of 20 minutes. At other times, such as between 7 pm and 7 am and on Sundays, unrestricted stopping is allowed in either type of loading box.

'Loading' is defined as when a vehicle stops briefly to load or unload bulky or heavy goods. These goods must be heavy or bulky enough so that they can't be carried any distance and may involve more than one trip. If possible your vehicle should be parked legally and the goods carried to the premises. Picking up items that are able to be carried, like shopping, doesn't constitute loading.

Clearways These are major roads where there's no need to stop. There will be no red lines, but Red Route clearway signs will indicate that stopping isn't allowed at any time.

For more information on Red Routes contact

Transport for London Street Management
Windsor House
50 Victoria Street
London
SW1H 0TL

Tel: 020 7941 4500
Webste: www.transportforlondon.gov.uk

Congestion charging

A congestion charging scheme was introduced in central London on 17 February 2003 to help reduce traffic and make journeys and delivery times more reliable. The congestion charge applies from 7.00 am to 6.30 pm Monday to Friday, excluding public holidays. Failure to pay the charge will lead to a fine.

Exemptions Those who are exempt from the charge include

- disabled drivers
- residents who live within the congestion charging zone
- alternative fuel vehicles
- vehicles with nine or more seats
- roadside recovery vehicles
- all two-wheeled vehicles
- London licensed taxis and minicabs.

LIMITS AND REGULATIONS

Drivers in some categories of exemption need to register with Transport for London (see congestion charging contact details below).

Discounts Businesses and other organisations operating a fleet of 25 or more vehicles are entitled to a discount when they register with a dedicated fleet scheme.

For more information, to register or to make a payment, ring the Congestion Charge Line, tel. 0870 900 1234 or visit the website: www.cclondon.com

Congestion charging is also in operation in Durham and may be introduced in other towns.

Bus lanes

Bus lanes are marked on busy roads to assist the flow of public transport. Traffic signs will indicate if coaches are permitted to use the lane. Use them sensibly and don't be tempted to speed just because the lane is clear ahead. You may be driving along the inside of stationary or slow-moving traffic where pedestrians could be tempted to cross the road. They may not be prepared for you moving faster along the bus lane.

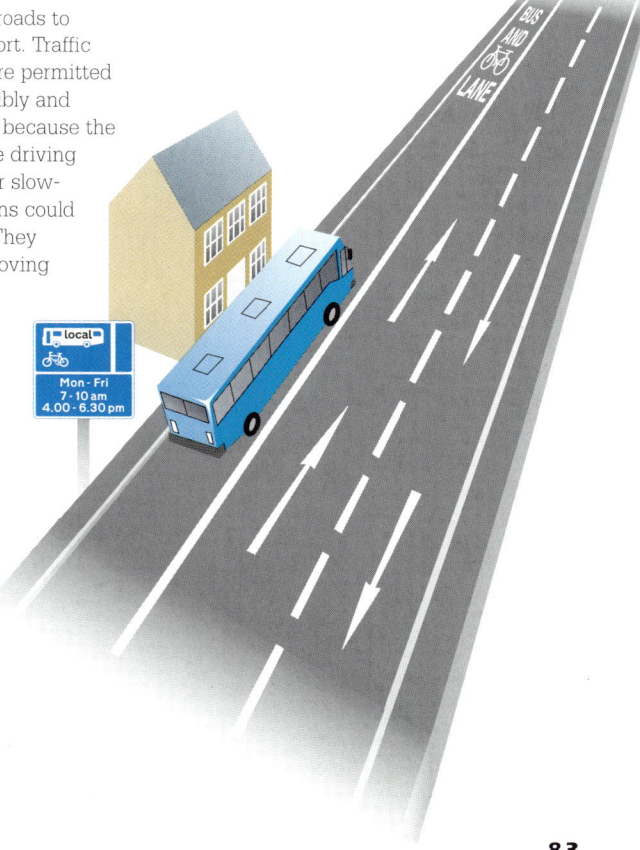

Where the lane has been obstructed, try not to get annoyed. It achieves nothing except to distract you from your driving.

Indicate in good time when you're ready to move out and wait patiently for an opportunity.

Be prepared for the end of the lane, where other traffic may be changing position.

83

LIMITS AND REGULATIONS

Parking restrictions

Whenever you're driving, whether on stage carriage, on a private hire contract, on tour work, or whilst not in service, maintain your professionalism. Don't stop in places where

- loading and unloading aren't allowed
- you'll cause an obstruction
- you'll inconvenience other road users.

Similarly, don't park

- where parking is prohibited
- where there's a risk of theft or vandalism
- less than 10 metres (33 feet) from any junction, wherever possible, unless there's an authorised parking space.

Route planning

Plan your route carefully. It's never advisable to take short cuts through residential or narrow streets to avoid traffic congestion; you may get stuck. In some towns weight, size and other vehicle restrictions prohibit you from leaving the main through-routes and ring roads, except for access.

The Metropolitan Police operate an advisory service for coach operators. Their telephone number is at the back of this book.

LIMITS AND REGULATIONS

Your passengers

Various regulations cover how you should deal with passengers and their behaviour. Specific rules relate to

- the conduct of drivers, conductors, couriers and inspectors
- the number of passengers carried
- the carriage of schoolchildren
- the carriage and consumption of alcohol
- smoking
- passengers causing danger or offence by their behaviour or condition
- the carriage or use of dangerous, noxious or illegal substances by passengers.

In addition to the legal obligations and restrictions, most operators require that specific rules must be followed. It's in your own interest to read and comply with them. You may risk dismissal if you don't.

LIMITS AND REGULATIONS

Health and Safety

A wide range of activities are covered by the Health and Safety regulations. These include

- limits on the weight of objects that should be lifted manually (for example, loading suitcases)
- requirements for protective clothing when handling oils and other maintenance materials, and when disposing of waste (emptying toilet tanks, etc.)
- safe operating procedures in the event of emergencies or breakdowns
- safe working practices in garages, bus depots and bus stations.

Safe working practices

Every year someone in the bus and coach industry is killed or badly injured in an incident involving moving vehicles in confined spaces. When parking close to a wall or another vehicle make sure that

- you leave room for other vehicles
- you're not trapping or crushing anyone.

The bodies of those vehicles fitted with air suspension may move a considerable amount when parked or when started, as air is exhausted or injected into the air bags. Parking one of these vehicles too close to a pillar, wall or another vehicle may cause damage or injury.

Vehicle maintenance and repair work isn't normally your responsibility. However, drivers are responsible for the condition of their vehicles when in use on the road. You should be able to recognise faults with your vehicle and fill in defect reports correctly, or inform the person responsible for recording faults. You may have to carry out minor emergency repairs on the road, when conditions dictate, but don't attempt anything beyond that. You shouldn't do any work on engines or any other vehicle components unless you're fully trained or supervised.

LIMITS AND REGULATIONS

Be careful of the following hazards in workshops and garages

- asbestos dust
- paint spray
- solvents
- exhaust fumes
- degreasing agents
- inspection pits
- moving/reversing vehicles
- vehicle batteries
- vehicle chair lifts or 'kneeling' mechanisms
- bus washers
- trailing cables or air lines
- spills of oil or fuel.

If you don't have to be in the workshop or garage, keep out.

The professional driving standards described in this book should also apply to drivers employed as 'shunters' or mechanics who drive buses and coaches as part of their job.

Anti-theft measures

There are many anti-theft systems on the market, some of which manufacturers are fitting as original equipment to vehicles.

This book isn't intended to provide a detailed description of the precautions that you can take to avoid having your vehicle stolen or broken into, except in general terms. To provide this information would only alert criminals to the ways in which they can be overcome.

Unless you're handing a vehicle over to another driver, or parking it on an operator's premises where it's safe to do so, don't

- leave a vehicle unlocked or unattended
- allow passengers to leave personal effects on board, except in locked luggage compartments
- forget to set any fitted anti-theft devices.

Remember, there have been numerous incidents when considerable damage has been done by buses or coaches which were driven away by unauthorised persons. Not only has there been damage to the stolen vehicle, but also to vehicles belonging to innocent parties.

The basic rules are simple

- avoid carelessness
- assess the risks of theft or damage
- set all devices fitted.

Part Four
Driving skills

The topics covered

- **Professional driving**
- **Driving at night**
- **Motorway driving**
- **All-weather driving**
- **Breakdowns**
- **Accidents**
- **First aid**

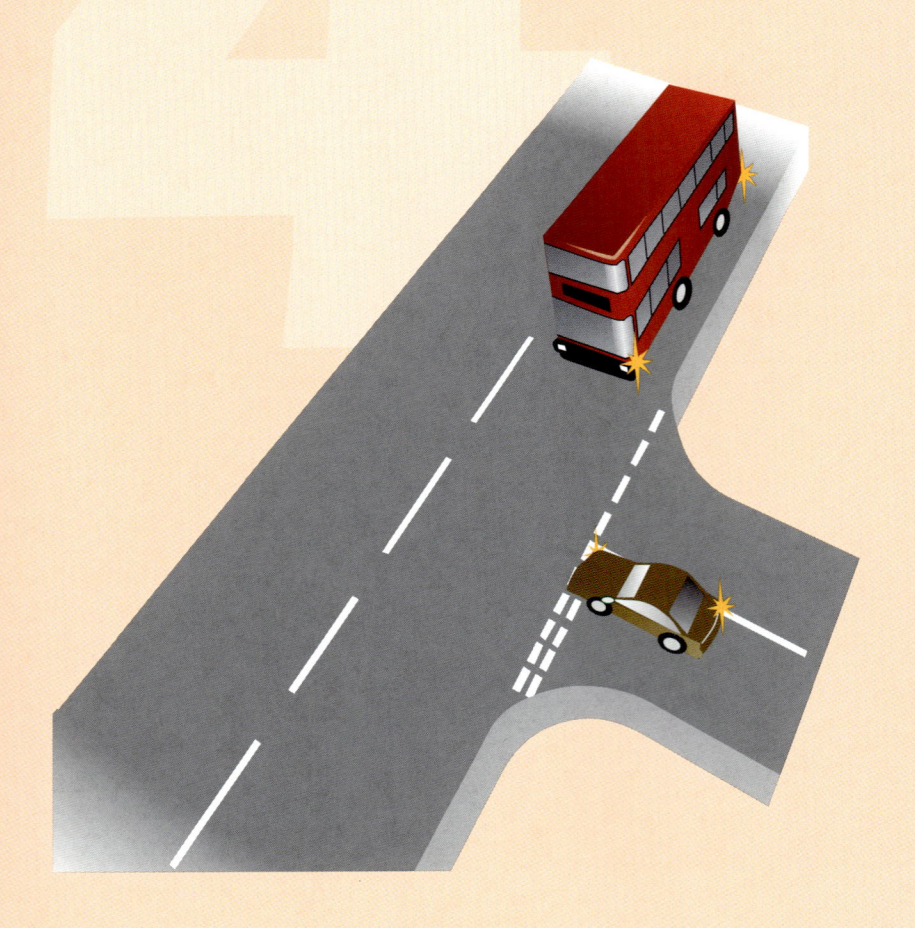

DRIVING SKILLS

Professional driving

Essential skills

Professional drivers adopt a positive approach to driving. This means

- looking after yourself, your vehicle and your passengers
- planning well ahead
- practising good observation
- keeping in control
- anticipating events.

Professional driving also means making allowances. You must always consider the safety and comfort of passengers. Sometimes you'll have to allow for the ignorance of other road users. In most cases, they'll have very little idea of the problems a bus or coach driver faces when driving such a large vehicle.

Control

It's essential that your vehicle is under control at all times. You must drive it skilfully and plan ahead, so that your bus is always travelling at the correct speed and ready for your next manoeuvre. You should never have to do anything at the last minute.

If you get caught out, you've got it wrong.

Awareness

You need to develop your awareness, to know what's going on round you at all times. This can be achieved through

- planning ahead
- anticipating – experience will soon tell you what other road users are probably going to do next
- being in control. Plan your actions, don't be forced into situations by others
- understanding what might happen
- remembering similar situations.

You must always drive

- responsibly
- carefully
- considerately
- courteously.

At all times, show that your standards are high and that you can drive a PCV with skill and safety.

89

DRIVING SKILLS

Anticipation

There aren't many excuses for being taken by surprise when you're driving; almost every event is predictable to some extent.

You must consider and prepare for all possibilities in all situations, especially when you can never be completely sure of what other road users will do. Remember, you won't be able to brake or swerve like lighter, smaller vehicles can.

Put yourself in other people's shoes. Make allowances for

- children
- cyclists
- horse riders
- elderly pedestrians
- obviously less able drivers
- learner drivers.

Problems particularly arise when you aren't sure of what vulnerable road users intend to do. Try to prepare yourself for all possibilities.

Avoiding aggression

Your passengers trust you; their safety is in your hands once they board your bus. Don't betray that trust. When you're driving

- accept that mistakes can be made
- expect others to make mistakes
- don't 'rise' to aggression.

People who drive aggressively often see their driving as a competition. It's preferable to let them go on their way. Refuse to be involved in their bad driving behaviour – and their accident.

DRIVING SKILLS

Your driving should always be a good example to others. By driving patiently and being prepared for the unexpected you'll avoid

- giving offence to others
- creating hostility
- provoking others to drive dangerously.

Safe procedure

It may seem to other road users that coach drivers are 'racing' when one coach overtakes another. This is usually because of a coach's load or its speed limiter, and tends to be more obvious on hills.

Sometimes the bus being overtaken may be more powerful and the overtaking vehicle must drop back. If another coach has started to overtake you but appears to be unable to pass, be prepared to adjust your own speed if you think that it would be safer for the other driver to move back to the left.

For this reason you must not drive in close convoy. If you're driving with other vehicles from the same company, don't drive nose-to-tail or look as though you're vying for position with each other along the road.

Competing with other drivers will eventually lead to you risking your own safety or that of your passengers and other road users.

DRIVING SKILLS

Effective observation

As a PCV driver, you'll often have a better view from your driving position than most other road users. You can take advantage of this, for example, when approaching a blind bend, by using your added height to see over hedgerows or other obstructions; you can then scan ahead for potential hazards. However, because of your vehicle's size and design, it will have more blind spots than many smaller vehicles.

You must use the mirrors effectively and act upon what you see in them. Just looking isn't enough. You need to know what road users around you are doing, or might do next. Check frequently down the sides of your vehicle.

Check the offside

- for overtaking traffic coming up behind, or already alongside
- before signalling
- before changing lanes, overtaking, moving or turning to the right.

Check the nearside

- for cyclists or motorcyclists 'filtering' up the nearside
- for traffic on your left when moving in two or more lanes
- to check when you've passed another road user, parked vehicle or pedestrians before moving back to the left
- to see where your wheels are in relation to the kerb or gutter
- before changing lanes, after overtaking, before turning left or moving further to the left, before leaving roundabouts.

You should use your mirrors frequently so that you're constantly aware of what's happening around you.

With a high seating position you must also be aware of pedestrians or cyclists. They may be out of sight below the windscreen line, directly in front of your vehicle. Check for them

- before moving off
- at pedestrian crossings
- in slow-moving congested traffic
- when manoeuvring to park.

Blind spots

In addition, inside some coaches – particularly those with high side windows – it's difficult to see to either side. When you want to move off you should open the window and look down and round to the right to ensure that it's clear before you pull away.

Many modern vehicles are fitted with additional mirrors on the left-hand side, positioned so that the driver can observe the nearside front wheel in relation to the kerb. Use them whenever you're pulling in to park alongside the kerb, in addition to checking the vehicle's position when you have to move close to the left in normal driving.

Hitting the kerb or wandering onto a verge can seriously deflect the steering or damage the tyre, which could result in a blow-out later.

DRIVING SKILLS

Observation at junctions

Despite your higher seating position there will still be some junctions where you can't see past parked vehicles or even road signs. If possible, try to look through the windows of other vehicles, or watch for other vehicles' reflections in shop windows, which can give you some valuable information.

If you still can't see properly you'll have to ease forward until you can do so, without emerging too far out into the path of approaching traffic. Remember, some road users are more difficult to see than others, particularly cyclists and motorcyclists.

- Lock
- Assess
- Decide before you
- Emerge or enter, then
- Negotiate the junction

If you don't know, don't go.

At junctions, check for everything that you would normally look for whenever you move off from a standstill position. For example, it can be difficult to predict what pedestrians might do at junctions. Sometimes they might run out into the road, or other times they might just step out without having seen you.

Never decide to go after just one quick glance. Take in the whole scene before you commit yourself to moving out.

Think once

Think twice

Think bike.

93

DRIVING SKILLS

Zones of vision

As a PCV licence-holder your eyesight must be of a high standard. A skilful driver will constantly scan the road ahead to see what's happening. You need to anticipate what might happen next.

Because you use the scanning technique you will know what's behind and next to you. Note what's happening at the edges of your vision and check what changes there are 'out of the corner of your eye'. You need to act on **all** your observations. Check for

- vehicles about to come out of junctions
- children running out
- bikes and motorcycles
- pedestrians stepping out.

Look for clues. If you see a cyclist ahead glance round to the right, they're probably going to try to turn right into the next road. Be ready for it. Similarly, watch the actions of pedestrians as they approach kerbs and cross the road. Elderly people sometimes become confused and change direction suddenly, or even turn back.

Keep a good look out for horse riders. If the animal starts to behave nervously allow the rider time and space to control it. The noise of a bus's exhaust or brakes can disturb even a normally calm horse.

DRIVING SKILLS

Safe distances

Never drive at such a speed that you can't stop in the distance that you can see is clear ahead. You need to do this regardless of the weather, the road and whether you're carrying passengers or not. This is one rule of safe driving that must **never** be broken.

- Keep a safe separation distance between you and the vehicle in front.
- In good weather conditions leave at least 1 metre (3 feet) per mph of your speed, or a two-second time gap.
- On wet roads you'll need to leave at least a four-second time gap.

The two-second rule

You can check the time gap easily. Watch the vehicle in front pass a stationary object such as a bridge, pole, sign, etc. and then say to yourself

'Only a fool breaks the two-second rule.'

You should have finished saying this by the time you reach the object. If you haven't, you're too close.

On some motorways this rule is drawn to drivers' attention by chevron markings painted on the road surface. The instruction 'Keep at least two chevrons from the vehicle ahead' also appears on a sign at these locations.

In busy, slow-moving traffic you may not need to leave as much space, but you must still leave enough distance in which to stop safely.

Tailgating

If you find another vehicle driving too close behind you, gradually ease your speed to increase the gap between you and any vehicle ahead. You'll then be able to brake more gently and remove the likelihood of the close-following vehicle running into you from behind.

If another vehicle pulls into the safe separation gap you're leaving, ease off your speed to extend the gap again.

Never drive, at speed, within a few feet of the vehicle in front. It isn't only car drivers in motorway right-hand lanes who commit this offence. Lorry and bus drivers can sometimes be seen driving much too close behind another vehicle – often at normal motorway speeds. If anything unexpected happens, an accident could follow.

You must not rely on someone else to plan ahead for you. They may not possess the same skills as you. Always keep your distance.

DRIVING SKILLS

Being aware of others

Look well ahead for stop lights. On a road with the national speed limit in force or on the motorway, watch for hazard warning lights flashing. These show that traffic ahead is slowing down quickly.

When you plan well ahead less effort is needed to drive a bus. You should be able to keep your vehicle moving by anticipating traffic speeds. Your fuel economy should improve and this could help your company to stay competitive.

Before you change direction or speed you must decide how any change will affect other road users. It's important to know what's happening behind you as well as what's going on in front of you. Fast-moving cars or motorcycles can catch up with you surprisingly quickly.

Bus or coach drivers can't usually see much by looking round, which is why you must always be aware of vehicles just behind you and to either your left- or right-hand side as they come into your blind spot position.

A quick sideways glance is often helpful, especially

- before changing lanes on a motorway or dual carriageway
- where traffic joins from the right or the left
- prior to merging from a motorway slip road.

Don't take your attention off the road ahead for any longer than is absolutely necessary.

Mirrors

You must use the mirrors well before you signal or make any manoeuvre, such as before

- moving away
- changing direction
- turning left or right
- overtaking
- changing lanes
- slowing or stopping
- speeding up
- opening any offside door.

Mirrors must be

- clean
- properly adjusted
- free from defects.

Whenever you use the mirrors you must act sensibly on what you see. Take note of the traffic behind you and what it's doing.

Just looking isn't enough.

Traffic lights

At many busy road junctions the road is covered in skid marks. This shows that vehicles have come up to the junction too fast and have had to brake hard.

Approaching traffic lights

Lights on green Ask yourself

- How long has green been showing?
- Can I stop safely at this speed if the lights change?
- If I do have to brake hard, will the traffic behind be able to stop safely?
- Are there any vehicles waiting to turn left or right?
- How will weather conditions affect my braking?

DRIVING SKILLS

Lights on red You must, of course, stop at red traffic lights. However, you may be able to time your approach so that you keep your vehicle moving as they change. Timing your approach to avoid stopping and moving off again may make your driving easier and your passengers more comfortable.

Lights not working If you come up to traffic lights that aren't working, or there's a sign to show that they're out of order, treat the junction like an unmarked junction and proceed with great care. Practise good, all-round observation and be prepared to stop if others assume priority.

Lights 'stuck' on red By law you must not go through a red traffic light, unless a police officer tells you to do so. Occasionally, the signals may go out of phase and the red light shows for longer than it should. Remember, if you drive on and there's an accident, you'll have broken the law.

Never attempt to 'beat' any traffic lights. Don't

- speed up to try to beat the signals. Remember what might happen to your passengers if you have to suddenly brake
- leave it until the last moment to brake. Heavy braking may well end up in loss of control.

A vehicle coming across your path may anticipate the lights changing and accelerate forward while the lights are still on red-and-amber. Don't take any risks.

Remember, a green light does not give you right of way, it means 'go on if the way is clear.' Check the junction to make sure other traffic using the junction stops at their red lights. Do not emerge at a green light if it will cause you to block the junction.

Giving signals

You should signal to

- warn others about what you're going to do, especially if this involves a manoeuvre that isn't obvious to other road users
- help other road users.

Road users you need to consider include

- drivers of oncoming vehicles
- drivers of following vehicles
- motorcyclists
- cyclists
- crossing supervisors
- police directing traffic
- pedestrians
- horse riders.

Give all signals clearly and in good time. Also, use only those signals that are shown in *The Highway Code*.

You should avoid giving any signals that could confuse, especially when you're going to pull up just past a road on the left. Another road user might misunderstand the meaning of the signal.

Avoid giving unauthorised signals, no matter how widely you assume they're understood. This applies to headlight 'codes' and alternating indicator signals. Remember, any signal that doesn't appear in *The Highway Code* is unauthorised and could be misunderstood by another road user.

Avoid unnecessary signals. Always consider the effect your signal will have on all other road users.

DRIVING SKILLS

Using the horn

There are few instances when you'll need to use the horn. Using it does not

- give you any 'right of way'
- relieve you of the responsibility of driving safely.

You should only sound the horn if you

- think that another road user may not have seen you
- need to warn other road users of your presence – at blind bends or a humpback bridge, for example

Don't use the horn

- to rebuke another road user
- simply to attract attention (unless to avoid an accident)
- when stationary (unless a moving vehicle presents a danger)
- at night between 11.30 pm and 7 am in a built-up area, unless there's danger from a moving vehicle.

Avoid long blasts on the horn, which can alarm pedestrians. If they don't react, they may be deaf.

Mobile phones

You must exercise proper control of your vehicle at all times. Never use a hand-held mobile phone or microphone when driving. Using hands-free equipment is also likely to distract your attention from the road. Find a safe place to stop before using all such equipment.

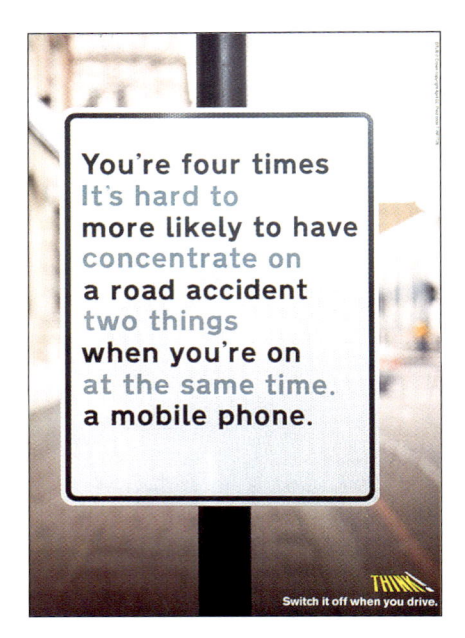

You're four times
It's hard to
more likely to have
concentrate on
a road accident
two things
when you're on
at the same time.
a mobile phone.

Switch it off when you drive.

DRIVING SKILLS

Driving through tunnels

On approaching and while in a tunnel

- switch on your dipped headlights
- do not wear sunglasses
- observe the road signs and signals
- keep an appropriate distance from the vehicle in front
- switch on your radio and tune to the indicated frequency.

If the tunnel is congested

- switch on your hazard warning lights
- keep your distance, even if you are moving slowly or stationary
- listen out for messages on the radio
- follow instructions given by tunnel officials or variable message signs.

If you break down or have an accident in a tunnel remember that your first priority is to the passengers on board your vehicle. Circumstances will dictate your actions, but in general

- switch on your warning lights
- switch off the engine
- if the vehicle is parked in a dangerous position then remove the passengers, keep them together and take them to the nearest exit point
- give first aid to any injured people, if you are able
- call for help from an emergency point.

If your vehicle is on fire, and you can drive it out of the tunnel, do so. If not

- pull over to the side and switch off the engine
- remove the passengers, keep them together and take them to the nearest exit point

- when the passengers are in a safe place, and without putting yourself in any danger, try to put out the fire using the vehicle's extinguisher or the one available in the tunnel
- move without delay to an emergency exit if you cannot put out the fire
- call for help from the nearest emergency point.

If the vehicle in front is on fire, switch on your warning lights, then follow the above procedure, giving first aid to any injured if possible.

99

DRIVING SKILLS

Driving at night

Problems encountered

You need extra skills to drive a bus, coach or minibus at night, especially over long distances. There are also additional responsibilities for the driver.

The problems related to driving at night are

- much less advance information
- limited lighting (street lights or vehicle lights only)
- dazzle from the headlights of oncoming vehicles
- shadows created by patchy street lighting
- poor lighting on other vehicles, pedal cycles, etc.
- dangers created by the onset of tiredness. Fatal accidents have happened because the driver of a large vehicle either fell asleep briefly or didn't see an unlit broken-down truck or car until it was too late.

You need to plan long journeys at night carefully, particularly on motorways where there's little to ease the boredom. You should also make sure that you get proper rest and refreshment stops.

Above all, you must drive at a speed that allows you to stop safely in the distance that you can see to be clear ahead. In many cases, that's within the distance illuminated by your headlights or by street lights.

DRIVING SKILLS

Tiredness

The smallest lapse of concentration at the wheel can result in loss of control. Many fatal accidents have been attributed to the driver becoming over-tired and falling asleep at the wheel. Remember

- don't commence your journey if you are tired
- don t drive without proper rest periods
- keep plenty of cool, fresh air circulating through the driving area
- don't allow the air around you to become too warm
- avoid eating a heavy meal before or during a journey
- pull up at the next safe, convenient place if you feel your concentration slipping
- listen to the radio or a tape if you can do so without disturbing your passengers (don't change tapes while driving, though).
- walk around in the fresh air before setting off again after a rest stop.

Night vision

Have your eyesight tested regularly and make sure that your night vision is up to the standard required. If in doubt, have it checked. Do not

- wear tinted glasses
- use windscreen or window tinting sprays.

Lighting-up time

Regardless of the official lighting-up times (when you must turn your lights on), you should be ready to switch on any lights that you may need. If the weather conditions are poor or it becomes overcast, don't be afraid to be the first driver to switch on.

See and be seen.

101

DRIVING SKILLS

Unlit vehicles

Only vehicles under 1,525 kg are allowed to park in 30 mph zones without lights at night time. Be on the alert when driving in built-up areas, especially when the street lighting is patchy.

Although builders' skips must be lit and show reflective plates to oncoming traffic, these items are often either forgotten or vandalised, so be on the lookout for skips.

Adjusting to darkness

When you step out from a brightly lit area into darkness, such as when leaving a motorway service station, your eyes will take a short while to adjust to the dark conditions. Use this time to check and clean your lights, reflectors, lenses and mirrors.

At dawn

Other drivers may have been driving through the night and may also be less alert. Leave your lights on until you're satisfied that other road users will see you.

Remember, it's harder to judge speed and distance correctly in the half-light at dusk and dawn. The colour of some vehicles makes them harder to see in half-light conditions. By switching your lights on you could avoid another road user stepping, riding or driving out into your path because they hadn't realised how close you were or how fast your vehicle was travelling.

See and be seen.

Vehicle lighting

It's essential that all lights are clean and that the bulbs and light units work properly. As well as being able to see ahead properly, other road users must be able to recognise the size of your vehicle and which way it's going.

In general, white lights indicate that the vehicle is

- moving towards you
- stationary, facing you
- reversing towards you (or is about to do so).

Red lights mean that the vehicle is

- moving away from you
- ahead of you and braking
- stationary, facing away from you.

Amber lights that aren't flashing mark the side of a vehicle.

Auxiliary lighting

High-intensity rear fog lights and additional front fog lights must only be used when visibility is less than 100 metres (about 330 feet). Remember to switch them off when conditions improve.

Interior lights

You should also turn on the interior lights if it's gloomy during the day, as well as at night. It helps passengers to move about the bus more easily and safely.

Coaches may have special lighting for night use. Never leave the interior of your coach in darkness when you have passengers aboard.

Interior lights have another role in road safety. Newer buses may have marker lights along the side to help make them more visible as they emerge from junctions, etc. Remember, a well-lit bus interior is even easier to see.

DRIVING SKILLS

Exterior lights

Buses and coaches which exceed 2.1 m wide and are first used on or after 1 April 1991 must be fitted with end-outline marker lamps to indicate the presence of a wide vehicle. The requirement is two white lamps to the front and two red lamps to the rear.

Parked vehicles

All buses and coaches and most minibuses – depending on their weight – must have lights on when parked on the road at night.

A lay-by is generally very close to the carriageway, and you must still have your lights on when parked in one. Unless your vehicle is parked 'off street', such as in a coach park, by law it must be clearly lit.

You must park on the left-hand side of the road unless you're on a one-way street and it's safe to park on the right-hand side of the road.

Driving in built-up areas

Always use dipped headlights in built-up areas at night. It helps others to see you and also aids your visibility if the street lighting changes or isn't working properly.

Watch out for

- pedestrians in dark clothing
- joggers
- cyclists (often without lights).

Take extra care when approaching pedestrian crossings. Drive at such a speed that you can stop safely if necessary.

Make sure that you still obey the speed limits even if the roads appear to be empty.

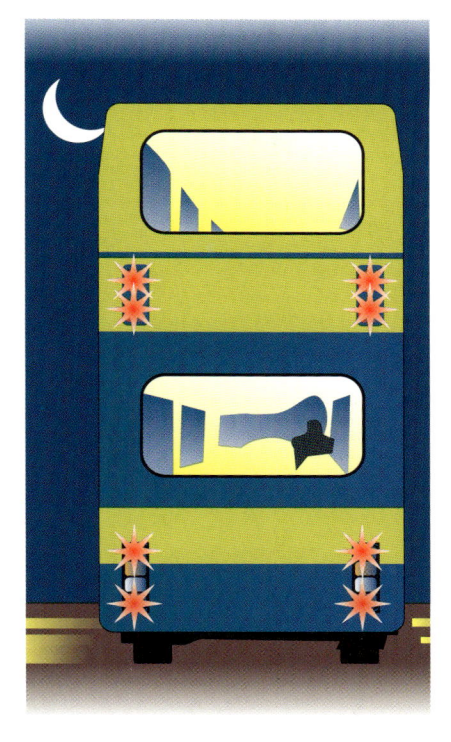

Maintenance work

Remember that essential maintenance work is often carried out at night time. Street cleansing in the larger cities often takes place at night, so be on the lookout for slow-moving vehicles.

Be on the alert for diversion signs, obstructions, coned-off sections of road, etc., which may be difficult to see at night.

103

DRIVING SKILLS

Driving in rural areas

If there's no oncoming traffic you should use full beam headlights to see as far ahead as possible. Dip your lights as soon as you see the lights of traffic coming towards you. This will avoid dazzling the oncoming driver or rider.

If there's no footpath, watch out for pedestrians on the nearside. *The Highway Code* advises pedestrians to walk facing oncoming traffic in these situations.

Be prepared for temporary traffic lights on rural roads.

Fog at night

If fog is forecast at night don't drive. You'll be a serious hazard to other traffic if the fog becomes so thick that you're unable to go any further safely. Because of the difficulties of getting a bus or coach off the road in thick fog it's better not to start out in the first place.

If you start your journey when there's fog about and you're delayed, you'll be committing an offence if you drive for more than your permitted hours. After all, the delay was foreseeable.

Seriously reduced visibility has, unfortunately, resulted in a number of major incidents involving multi-vehicle pile-ups. If conditions become severe enough, scheduled journeys may have to be cancelled. There's ample justification for putting caution before inconvenience.

DRIVING SKILLS

Overtaking at night

Because PCVs can take longer to overtake other vehicles, you must only attempt to overtake when you can see well ahead that it's safe to do so.

This means that, unless you're driving on a dual carriageway or motorway, you'll have few opportunities to overtake. Unless there's street lighting, you might not be able to see properly if there are bends, junctions, hills, etc., which may prevent you from seeing an oncoming vehicle.

If you do decide to overtake, make sure that you can do so without 'cutting in' on the vehicle you're overtaking, or causing oncoming vehicles to brake or swerve. Never close on the vehicle ahead before you attempt to overtake it, as this will restrict your view ahead.

When overtaking, switch to main beam headlights when you're past the point that you would dazzle the driver in their external mirrors. Using headlights will improve your vision ahead, but do not dazzle approaching traffic (on a dual carriageway, for instance).

Separation distance

Avoid driving so close to the vehicle in front that your lights dazzle the other driver. Make sure that your lights are on dipped beam.

If another vehicle overtakes you, dip your headlights as soon as the vehicle starts to pass you. Your headlight beam should fall short of the vehicle in front.

DRIVING SKILLS

Motorway driving

Accident records show that, statistically, motorways are the safest roads in the UK. However, motorway accidents often involve several fast-moving vehicles and consequently result in more serious injuries and damage than accidents on other roads.

Because of the high numbers of large vehicles using motorways many of these accidents involve lorries and, occasionally, coaches and minibuses. But if everyone who used the motorway drove to the same high standard as PCV drivers, it's arguable that many of these incidents could be avoided.

There's often little room for error when driving at speed on a motorway. The generally higher speeds and the volume of traffic mean that conditions can change much more quickly on motorways than on other roads. Because of this you need to be

- totally alert
- physically fit
- concentrating fully
- assessing well ahead.

If you aren't, you may fail to react quickly enough to any sudden change in traffic conditions.

DRIVING SKILLS

Fitness

Don't drive if you're

- tired
- feeling ill
- taking medicines that could affect your driving
- unable to concentrate for any reason.

Any of these factors could affect your reactions, especially in an emergency.

Rest periods

You must take the compulsory rest periods in your driving schedule. On long journeys, try to plan them to coincide with a stop at a motorway service area. This is especially important at night, when a long journey can make you more tired than usual.

If you eat a large meal immediately before driving, the combined effects of a warm coach the constant drone of the engine and long, boring stretches of road, especially at night, can soon cause the onset of drowsiness. Falling asleep at the wheel can happen so easily; don't let it happen to you.

It's against the law to stop anywhere on the motorway, including the hard shoulder and slip roads, for a rest. Tiredness is foreseeable and isn't considered to be an emergency.

If you start to feel even slightly tired, open the windows, turn the heating down and get off the motorway at the next junction. Even if you aren't scheduled to stop, it's preferable to falling asleep at the wheel.

When you reach a service area have a hot drink, wash your face (to refresh yourself) and walk round in the fresh air before driving on.

Regulations

You must follow the special motorway rules and regulations. Study the sections in *The Highway Code* that relate to motorways. Know, understand and obey any warning signs and signals.

DRIVING SKILLS

Vehicle checks

Before driving on the motorway you should ensure that you carry out routine checks on your vehicle, especially considering the long distances and prolonged higher speeds involved.

For fuller details on vehicle maintenance, see page 46.

Tyres

All tyres should be checked regularly, especially if you are going to drive on the motorway. Tyres can become hot and may disintegrate under sustained high speed running (for tyre care and maintenance see pages 50-51).

Mirrors

Ensure that all mirrors are properly adjusted to give the best possible view to the rear. Also, make sure that they're clean.

In winter, make full use of any demisting heating elements fitted to your mirrors.

Keep the lenses and screens of any rear-view video equipment clean and clear.

DRIVING SKILLS

Windscreen

All glass must be

- clean
- clear
- free from defects.

Keep all windscreen washer reservoirs topped up and the jets clear. Make sure that all wiper blades are in good condition.

Don't hang mascots or put stickers where they could restrict your view or distract you.

Spray-suppression equipment

It's essential that you check all spray-suppression equipment fitted to the vehicle before setting out, especially if bad weather is expected. If wheel arches have sections of anti-spray fitments missing, report it as a defect.

Instruments

Check all gauges, especially any warning lights – air, oil pressure, coolant and the like.

Lights and indicators

By law, all lights must be in working order, even in daylight. Make sure that all bulbs, headlight units, lenses and reflectors are fitted, clean and working properly.

High-intensity rear fog lights and marker lights (if fitted) must also work correctly. Indicator lights must flash between 60 and 120 times per minute. Reversing lights must either work automatically when reverse gear is chosen or be switched on from the cab, with a warning light to show when they're on.

DRIVING SKILLS

Fuel

Make sure that you either have enough fuel on board to complete the journey or have the facility (cash, agency card, etc.) to refuel at a service area.

Oil and coolant

The engine operates at sustained high speeds on a motorway so it's vital to check all oil levels before setting out. Running low can result in costly damage to the engine and could cause a breakdown at a dangerous location. Similarly, it's essential to check the levels of coolant in the system.

Audio and video equipment

Don't allow the use of such equipment to distract you from driving carefully and safely. You shouldn't use microphones or headphones of any kind, or try to tune the radio whilst driving. If your bus is fitted with a communications radio or telephone, you should only use it whilst driving if it's fitted with a hands-free microphone. Otherwise find a safe place to stop before using a hand-held one.

Any video or television screen fitted to your coach for passenger use must not be visible to you while you're driving.

DRIVING SKILLS

Joining a motorway

There are three ways in which traffic can join a motorway. All these entrances will be clearly signed.

At a roundabout

The motorway exit from a roundabout will be signposted to prevent traffic that doesn't want to use the motorway from driving onto it unintentionally.

Main trunk road becoming a motorway

There will be prominent advance warning signs so that prohibited traffic can leave the road before the motorway regulations come into force.

Via a slip road

Slip roads leading directly onto the motorway will be clearly signed to prevent prohibited traffic entering the motorway. In many cases the slip road begins as an exit from a roundabout.

Effective observation

Before joining the motorway from a slip road, try to assess the traffic conditions on the motorway itself. You may be able to do this as you approach from a distance or if, before joining it, you have to cross the motorway by means of a fly-over.

Get as much advance information as you can to help you to plan your speed on the slip road before reaching the acceleration lane. You must give way to traffic already on the main carriageway. Plan your approach so that you don't have to stop at the end of the acceleration lane. Never use the size or speed of your vehicle to force your way onto the motorway.

Use the MSM/PSL routine. A quick sideways glance may be necessary to ensure that you correctly assess the speed of any traffic approaching in the nearside lane. Remember to

- look for approaching traffic
- assess the speed of approaching vehicles
- decide when you can build up speed
- emerge safely onto the main carriageway
- negotiate the hazard – adjust to the speed of traffic already on the motorway.

Don't

- pull out into the path of traffic in the nearside lane if this would cause it to slow down or swerve
- drive along the hard shoulder to 'filter' into the left-hand lane.

At a small number of locations traffic merges onto the motorway from the right. Take extra care in this situation.

111

DRIVING SKILLS

Making progress

Approaching access points

After passing a motorway exit there will usually be an entrance onto the motorway. Look well ahead and, if there are several vehicles joining the motorway

- don't try to race them while they're in the acceleration lane
- be prepared to adjust your speed
- move to the next lane, if it's safe to do so, to allow joining traffic to merge.

Lane discipline

Keep to the left-hand lane unless you're overtaking slower vehicles. The right-hand lane of a motorway with three or more lanes MUST NOT be used (except in prescribed circumstances) if you are driving

- any vehicle with a trailer
- a passenger vehicle with a maximum laden weight exceeding 7.5 tonnes that is constructed or adapted to carry more than eight seated passengers in addition to the driver, unless there are roadworks and signs directing otherwise.

This restriction does not apply to A class roads with three lanes.

On two-lane motorways, all vehicles may use the right-hand lane for overtaking.

If a bus or coach is fitted with a speed limiter it will be set to a max speed of 100 kph (62 mph), so consider this before attempting any overtaking.

Watch out for signs showing a crawler/climber lane for LGVs. This will suggest a long, gradual gradient ahead.

Use the MSM/PSL routine well before you signal to move out. Don't start to pull out and then signal, or signal at the same time as you begin the manoeuvre. Other drivers need time to react.

On a three- or four-lane motorway make sure that you check for any vehicles in the right-hand lane or lanes that might be about to move back to the left. Most of the traffic coming up from behind will be travelling at a higher speed. Look well ahead to plan any overtaking manoeuvre, especially given the effect a speed limiter will have on the power available to you.

If a very large slow-moving vehicle is being escorted, watch for any signal by the police officers in the escort vehicle at the rear. You might need to move into the right-hand lane to pass it.

If a motorway lane merges from the right (this only happens in a few places) you should move over to the left as soon as it's safe to do so. The MSM/PSL routine must be used, with careful checks in the left-hand mirror and constant awareness of vehicles in the blind spots.

DRIVING SKILLS

Separation distance

When driving at motorway speeds you must allow more time for everything that you do. Allow

- greater safety margins than on normal roads
- a safe separation distance.

In good conditions that means you'll need at least

- 1 metre (about 3 feet 3 inches) for every mph
- a two-second time gap.

In poor conditions you'll need at least

- double the distance
- a four-second time gap.

In snow or icy conditions the stopping distances can be up to ten times those needed in normal dry conditions.

Seeing and being seen

Make sure that you start out with a clean windscreen, mirrors and windows. Use the washers, wipers and demisters to keep the screen clear. In poor conditions use dipped headlights.

Keep reassessing traffic conditions around you. Watch out for brake lights or hazard warning lights that show the traffic ahead is either stationary or slowing down. (Hazard warning lights may be used on moving vehicles to alert traffic to danger ahead.)

High-intensity rear fog lights must only be used when visibility falls below less than 100 metres (about 330 feet). They should be switched off when visibility improves, unless fog is patchy and danger still exists.

113

DRIVING SKILLS

Motorway signs and signals

Motorway signs are larger than most normal road signs. They can be read from further away and can help you to plan ahead.

Know your intended route. Be prepared in good time for the exit you require.

Where there are major roadworks there may be diversions for large vehicles. Look for the yellow

- square
- diamond
- circle
- triangle

symbols and follow the symbol on the route signs.

Signals

Warning lights show when there are dangers ahead such as

- accidents
- fog
- icy roads.

Look out for variable-message warning signs, which will warn you about

- lane closures
- speed limits
- hazards
- traffic stopped ahead.

Red light signals

If the red X signals show on the gantries above your lane, don't go any further in that lane.

- Be ready to change lanes.
- Be ready to leave the motorway.
- Watch out for brake lights and hazard warning lights showing that traffic has stopped or is moving very slowly ahead.

If the matrix sign indicating 'Stop, all lanes ahead closed' shows over every lane, stop and wait. You may not be able to see the reason for the signals and other drivers may be ignoring them. Remember, you're a professional driver who should know what the signals mean and can demonstrate to other drivers what they should do.

React in good time.

DRIVING SKILLS

Weather conditions

Because of the higher speeds on motorways, it's important to remember the effects that the weather can have on driving conditions.

Crosswinds

Be aware of the effects of strong crosswinds on other road users.

In particular, watch out for these effects

- after passing motorway bridges
- on high, exposed sections of road
- when passing vehicles towing caravans, horse boxes, etc.

If you're driving a high-sided vehicle, such as a double-decker or a high-floor coach, take notice of the warnings for drivers of such vehicles. Avoid known problem areas such as viaducts and high suspension bridges, if possible.

Motorcyclists are especially vulnerable to severe crosswinds on motorways. Watch out for them. Allow plenty of room when overtaking, and check the left-hand mirror after you've overtaken them.

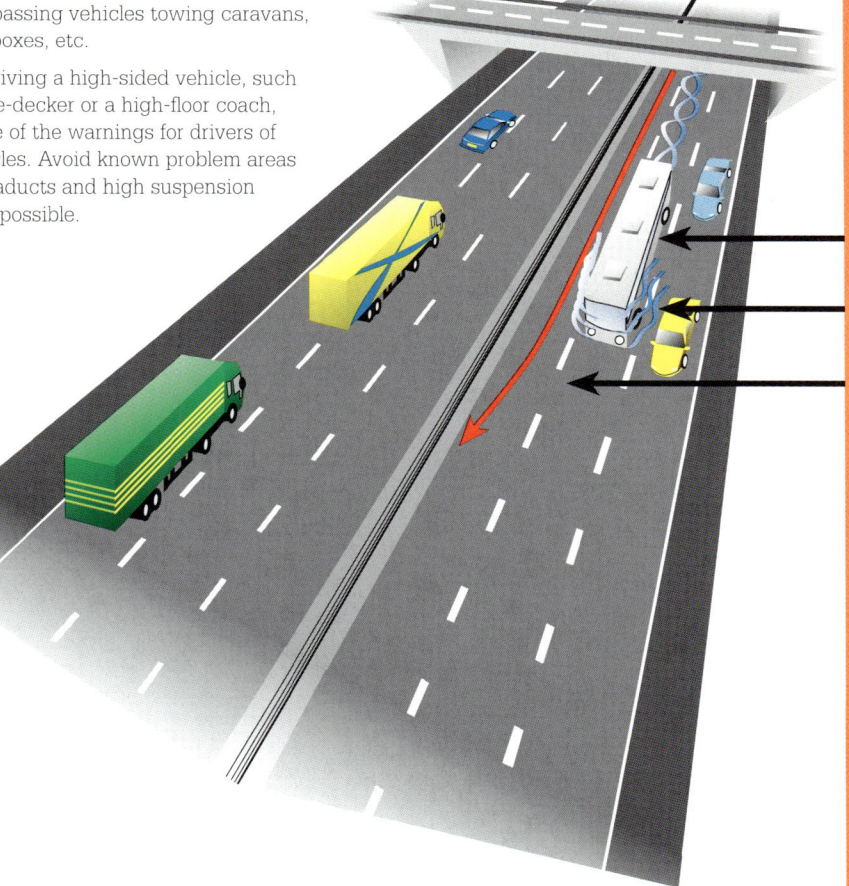

115

DRIVING SKILLS

Rain

The spray thrown up by large, fast-moving vehicles can make it very difficult to see ahead.

- Use headlights so that other drivers can see you.
- Reduce speed when the road surface is wet. You need to be able to stop in the distance that you can see is clear.
- Leave a greater separation gap. Remember the four-second rule as a minimum.
- Make sure that all spray-suppression equipment fitted to your vehicle is working.

Take extra care when the surface is still wet after rain. Roads can still be slippery even if the sun's out.

Ice or frost

In cold weather, especially at night when temperatures can drop suddenly, watch out for any feeling of 'lightness' in the steering (not always obvious with power steering). This may suggest frost or ice on the road. Watch for signs of frost along the hard shoulder. Remember, a warm coach interior can isolate you from the real conditions outside.

Motorways that appear wet may in fact be frozen. There are devices that fix onto an outside mirror to show when the temperature drops below freezing point. Also, some manufacturers fit ice-alert warning lights on the instrument panel.

Allow up to **ten** times the normal distance for braking in these conditions. And remember, all braking must be gentle.

DRIVING SKILLS

Fog

If there's fog on the motorway you must slow down so that you can stop in the distance that you can see is clear. You should

- use dipped headlights
- use the rear high-intensity fog lights if visibility is less than 100 metres (about 330 feet)
- stay back
- check your speedometer.

Don't

- speed up again if the fog is patchy; you could quickly run into dense fog again
- hang onto the rear lights of the vehicle in front.

Fog affects your judgement of speed and distance. You may be travelling faster than you think.

Slow down.

Multiple pile-ups on motorways don't just happen – they're caused by drivers who

- travel too fast
- drive too close
- assume nothing has stopped ahead
- ignore signals.

You can't see well ahead in fog.

Watch out for any signals that tell you to leave the motorway. Also, look for accidents ahead and for emergency vehicles coming up behind (possibly on the hard shoulder). Police cars may be parked on the hard shoulder with their lights flashing. This might mean that traffic has stopped on the carriageway ahead.

'Motorway madness' is the term used to describe the behaviour of those reckless drivers who drive too fast for the conditions. The police prosecute drivers after serious multiple accidents. This is to get the message across to all drivers that they must

slow down in fog.

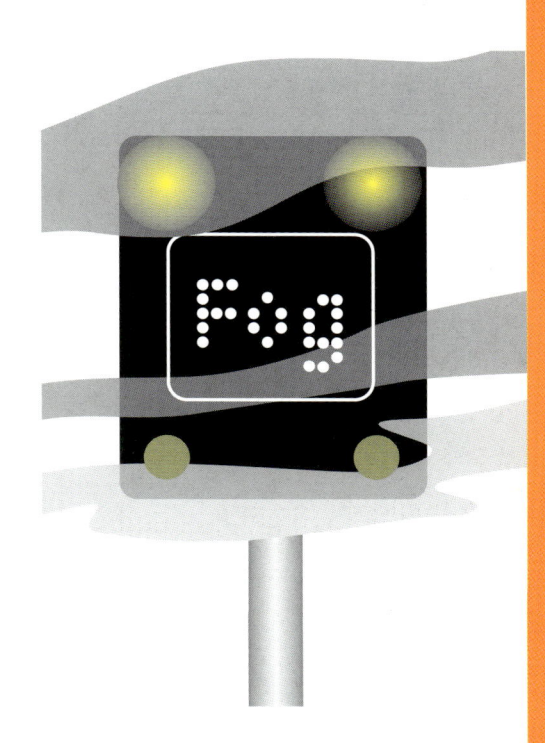

DRIVING SKILLS

Contraflows and roadworks

Essential roadworks involving two-way traffic on one carriageway of the motorway are called contraflow systems. The object is to let traffic carry on moving while repairs or resurfacing take place on the other carriageway or lanes.

Red and white marker posts are used to separate opposite streams of traffic. The normal white lane-marking reflective studs are replaced by temporary yellow/green fluorescent studs.

A 50 mph compulsory speed limit is usually in force in contraflow systems. Thus, in the event of a head-on collision the closing speed will be about 100 mph.

When driving through roadworks or in a contraflow, you should

- concentrate on what's happening ahead
- keep a safe separation distance from the vehicle in front
- look well ahead to avoid having to brake hard
- obey advance warning signs that tell you which lanes must not be used by large vehicles
- avoid sudden steering movements or any need to brake sharply.

Don't

- let the activity on the closed section distract you
- break the speed limit
- change lanes if signs tell you to stay in your lane
- speed up until the end of the roadworks and normal motorway speed limits apply again.

Accidents

Serious accidents can occur when vehicles cross into the path of the other traffic stream in a contraflow. You must

- keep your speed down
- keep your distance
- stay alert.

Signs

Take notice of advance warning signs (often starting five miles before the roadworks). Get into the correct lane in good time and don't force your way in at the last moment.

Breakdowns

If your vehicle breaks down in the roadworks section, stay with it. These sections of motorway are usually under TV monitoring, so a recovery vehicle (free within the roadworks section) will be with you as soon as possible.

Watch out for broken-down vehicles blocking the road ahead.

DRIVING SKILLS

Breakdowns on the motorway

If your vehicle develops a problem, leave the motorway at the next exit or pull into a service area. If you can't do this, pull onto the hard shoulder and stop as far to the left as possible.

- Switch on the hazard warning lights.
- Make sure that the vehicle lights are on at night, unless there's an electrical problem.
- Ensure passengers keep away from the carriageway and hard shoulder, and that children are kept under control.
- Move passengers as far forward in the vehicle as possible. This should help to limit injuries if another vehicle runs into the back of it.
- Don't try to carry out even minor repairs on the motorway.

You must not place a warning triangle or any other warning device on the carriageway, hard shoulder or slip road.

Emergency telephones

Motorway emergency telephones are free and easily located. You'll be connected directly to the motorway police control centre, who will then get in touch with a recovery company for you.

In most cases the emergency telephones are 1.6 km (about 1 mile) apart. The direction of the nearest phone will be shown by the arrow on the marker posts along the edge of the hard shoulder. Don't cross the carriageway or any slip road to get to a telephone. Face the oncoming traffic while using the telephone.

If your vehicle has its own telephone, make sure that whoever you contact also informs the motorway police, or telephone them yourself.

If you use a mobile phone, identify your location from the marker posts on the hard shoulder before you phone.

If anything falls from either your vehicle or another vehicle

- use the nearest emergency telephone to call the police
- don't attempt to recover it yourself
- don't stand on the carriageway to warn oncoming traffic.

119

DRIVING SKILLS

Leaving the motorway

Progressive signs will show upcoming exits. At one mile you'll see the

- junction number
- road number
- one-mile indicator.

Half a mile from the exit you'll see signs for the

- main town or city served by the exit
- junction number
- road number
- half-mile indicator.

Finally, from 300 yards (270 metres) before the exit there will be three countdown markers, one every 100 yards.

Remember, the driver of a vehicle travelling at 60 mph has 60 seconds from the one-mile sign to the exit. Even at a speed of 50 mph there's still only 80 seconds from the one-mile sign to the exit.

Plan well ahead in order to be in the left-hand lane in good time. Large vehicles in the left-hand lane may prevent a driver in the second lane from seeing the one-mile sign, leaving very little time to move to the left safely.

You must use the MSM/PSL routine in good time before changing lanes or signalling. Assess the speed of traffic well ahead. Avoid the situation where you try to overtake but then have to pull back in quickly in order to slow down to leave the motorway at the next exit. Don't

- pull across the carriageway at the last moment
- drive over the white chevrons that divide the slip road from the main carriageway.

If you miss the exit that you wanted to take, drive on to the next one.

Occasionally there are several exits close together or a service area close to an exit. Look well ahead and plan your exit in good time. Watch out for other drivers' mistakes, especially those who leave it too late to exit from the motorway safely.

Traffic queuing

In some places traffic can be held up on the slip road. Look well ahead and be prepared for this. Don't queue on the hard shoulder.

Illuminated signs have been introduced at a number of these sites to give advance warning of traffic queuing on the slip road or in the first lane. Watch out for indicators and hazard warning lights when traffic is held up ahead.

Use the MSM/PSL routine in good time and move to the second lane if you aren't leaving at such an exit.

DRIVING SKILLS

End of the motorway

There are 'End of motorway regulations' signs

- at the end of slip roads
- where the road becomes a normal main road.

Reduce speed

After driving on the motorway for some time it's easy to become accustomed to the speed. When you first leave the motorway, 40 or 45 mph seems more like 20 mph. You should

- adjust your driving to the new conditions as soon as possible
- check the speedometer to see your actual speed.

Start reducing speed when you're clear of the main carriageway. Remember, motorway slip or link roads often have sharp curves that need to be taken at lower speeds.

Look well ahead for traffic queuing at a roundabout or traffic signals. Be prepared for the change in traffic at the end of the motorway. Watch out for pedestrians, cyclists etc.

These remind you that different rules apply to the road that you're joining. Watch out for signs advising you of

- speed limits
- dual carriageways
- two-way traffic
- clearways
- motorway link roads
- part-time traffic signals.

121

DRIVING SKILLS

All-weather driving

Passengers want to travel 24 hours a day, 365 days a year. Whatever the weather, you'll need to drive safely so that you, your passengers and your vehicle arrive safely at your destination, with as few hold-ups as possible.

It is essential that you take notice of warnings of severe weather such as

- high winds
- floods
- fog
- snow and blizzards.

If a bus or coach becomes stranded the road may become blocked for essential rescue and medical services. In the case of fog it could result in other vehicles colliding with the stranded vehicle.

Training and preparation are vital. Don't venture out in severe conditions without being properly prepared. If the weather is really bad, cancel or postpone your journey.

Your vehicle

Your vehicle must be in good condition at all times. This means regular safety checks and strict observance of maintenance schedules. Far too many cases brought before a Traffic Commissioner result from incidents caused by a vehicle that wasn't looked after properly. Make sure that the vehicle you drive is fully roadworthy. For fuller details on vehicle maintenance, see page 46.

Tyres

Check the tread depth and pattern. Examine tyres for cuts, damage or signs of cord visible at the side walls.

Brakes

It's essential that the brakes are operating correctly. This is especially important on wet, icy or snow-covered roads. Any imbalance could cause a skid if the brakes are applied on a slippery surface.

Oil and fuel

Use the correct grades of fuel and oil in very hot or very cold weather.

Long periods of hot weather will make the oil in engines and turbo-chargers work harder. You should always allow engines with turbos fitted to idle for about a minute before increasing engine revs above tick-over speed (when starting) or before stopping the engine. This prevents the bearings from being starved of oil.

In extremes of cold you'll have to use either diesel fuel 'anti-waxing' additives or a suitable grade of diesel fuel with these properties to stop fuel lines freezing up.

DRIVING SKILLS

Use of the correct coolant when topping up prevents dilution of the rust inhibitors and antifreeze components of the liquid. Also, remember that allowing a cooling system to freeze will wreck components and possibly crack the engine block or cylinder heads.

Icy weather

Ensure that the whole of the windscreen is clear before you drive away in frosty conditions. Make full use of all heaters and demisters fitted.

If you're driving at night, remember that falling temperatures may lead to icy conditions. This will cause ungritted roads to become very slippery. If the steering feels light you're probably driving on ice. Ease your speed as soon as you can. All braking must be gentle and over much longer distances.

Leave more time for the journey, because you'll need to drive more slowly than usual. On slippery surfaces, keep a safe separation distance from any vehicle in front. Allow **ten** times the normal stopping distance.

Drive sensibly, and be careful of other road users getting into difficulties. Don't accelerate, brake or steer suddenly. No risks are ever justified.

If conditions are really bad, don't drive.

Heavy rain

You must make sure that you can see clearly ahead at all times. Don't drive if a windscreen wiper is faulty, even though many PCVs have more than one pair of wipers. In addition, the windscreen must be demisted fully and the windscreen washer bottle(s) topped up with the correct washing fluid. This is particularly important in winter. It's against the law to drive with frozen or ineffective windscreen washers.

Allow more space for braking – at least twice as much as in dry conditions. If possible, brake only when the vehicle is stable (and preferably when travelling in a straight line). Also, avoid sudden or hard braking.

Obey advisory speed limit signs on motorways.

Other road users will have more difficulty seeing when there's heavy rain and spray so make sure that all spray-suppression equipment on your vehicle is secure and working correctly.

Don't use rear fog lights unless visibility is less than 100 metres (330 feet). Fog lights reflect and dazzle following drivers.

123

DRIVING SKILLS

Mud

Take care when driving on off-road sites, such as at rallies, showgrounds, festivals, etc. If you get stuck there's very limited scope to 'rock' your PCV out of ruts (as you might with a car). Clearance underneath is often so limited that the exhaust could be ripped off, even if the vehicle has sunk as little as four or five inches. Seek assistance before this happens.

A bar from another vehicle on hard standing or the use of a winch may recover the situation without damage. Once the vehicle has sunk in, however, only the use of heavy-duty jacks and steel sheeting will get you out without further problems.

If the surface is hard but slippery, drive at a crawling pace in the highest possible gear and with the minimum 'revs'. Try to select a course that avoids you having to turn.

It's against the law to deposit mud on the road to the extent that it could endanger other road users.

Snow

Falling snow can reduce visibility quite seriously. Use dipped headlights and slow down. Leave a much greater stopping and separation distance – up to ten times the stopping distance on dry roads.

Road markings and traffic signs can be covered by snow. Take extra care at junctions.

High-level or exposed roads are sometimes closed by deep snow. Listen to any weather warnings. Don't try to use such roads if

- warning signs indicate that the road is closed to large vehicles or other traffic
- severe weather is forecast.

Some country roads in exposed places have marker posts at the side of the road, which tell drivers how deep the snow is. Remember, if a bus gets stuck it could

- stop snow ploughs from clearing the road
- delay emergency vehicles
- cause other road users to become stuck
- put passengers at risk.

If your vehicle is fitted with a manually selected retarder system, engage it before going down a hill covered with snow.

124

DRIVING SKILLS

Another technique for freeing a vehicle stuck in snow is to use the highest gear possible to try to get out. Alternating between the reverse and forward gears, if possible, is a good way of getting moving again when the snow is soft. Don't keep revving in a low gear; you'll only make the driving wheels dig in even further.

It's often helpful to keep a couple of strong sacks in your vehicle to put under the drive wheels if you get stuck, but remember the warnings about your vehicle undercarriage. Hardened snow can cause considerable damage. A shovel is often handy if you must go through areas where snow is a problem during the winter.

Make sure that your vehicle is properly prepared for any journey. This is especially important in the winter. In some countries you must carry snow chains at certain times of the year, and they must be used in bad weather.

Ultimately, ask yourself whether your route should go through an area where such conditions are likely.

No risk is worth taking.

Ploughs and gritting vehicles

Don't try to overtake a snow plough or gritting vehicle. You may find yourself running into deep snow or skidding on an ungritted stretch of road, which these maintenance vehicles could have treated had you followed on behind them.

Keep well back from gritters. Their presence could mean that the weather is already bad or that it's expected to be.

DRIVING SKILLS

Fog

Don't drive in dense fog if you can postpone your journey and avoid driving at all at night if there is fog.

Don't start a journey that might need to be abandoned because it becomes too dangerous to drive any further.

If you must drive in fog, **slow down**.

Also, keep a safe separation distance from any vehicle in front. If you can see the rear lights of a vehicle in front you're probably too close to stop in an emergency.

A large vehicle ahead of you may temporarily displace some of the fog, making it seem thinner than it really is. Overtaking at that point could quickly lead to a problem. Stay back.

Then again, in a larger vehicle you may be able to see ahead over low-lying fog. Don't speed up in case there are smaller vehicles in front that may be hidden from view. Only overtake if you're sure that the road ahead is clear – and then only on a multi-lane road.

Slow down

- Don't speed up if the fog appears to thin. It could be patchy and you may run into it again.
- Keep checking the speedometer to see your true speed. Fog can make it difficult to judge speed and distance.

Stay back

- Keep a safe separation distance from any vehicle ahead.
- Don't speed up if a vehicle appears to be close behind.
- Only overtake if you can be sure the road ahead is clear.

Don't take risks.

There aren't many places where you can find a safe place to park a bus in thick fog. You must not leave a bus on or near a road where it could be a danger to other road users. And your passengers won't be pleased with the prospect of spending the night in a lay-by. Certainly don't park a bus anywhere in fog without lights.

Lights

Use dipped headlights whenever you find it difficult to see. You need to see clearly and be seen at all times.

Use high-intensity rear fog lights and front fog lights (if fitted) when visibility is less than 100 metres (330 feet). Rear fog lights must only be capable of operating with dipped headlights or front fog lights. Switch off front and rear fog lights when you can see further than 100 metres (330 feet), but beware of patchy fog.

Keep all lights and reflectors clean and make sure that they're working correctly at all times, particularly in bad weather.

DRIVING SKILLS

Reflective studs

Reflective studs are provided on dual carriageways and motorways to help drivers to see in poor visibility. The colours of reflective studs are

- **red** – on the left-hand edge of the carriageway
- **white** – to indicate lane markings
- **amber** – between the right-hand edge of the carriageway and the centre reservation
- **green** – at slip roads and lay-bys
- **yellow/green** fluorescent – at roadwork contraflow systems.

On some country roads there are black and white marker posts with red reflectors on the left-hand side and white reflectors on the right-hand side of the road.

All these reflective devices are designed to help you know where you are on the road.

In fog don't

- drive too close to the centre of the road
- confuse centre lines with lane markings
- drive without using headlights
- use full beam, especially when following another vehicle. You'll make it more difficult for the other driver to see by casting shadows and causing glare in the mirror.

Slow down and stay back.

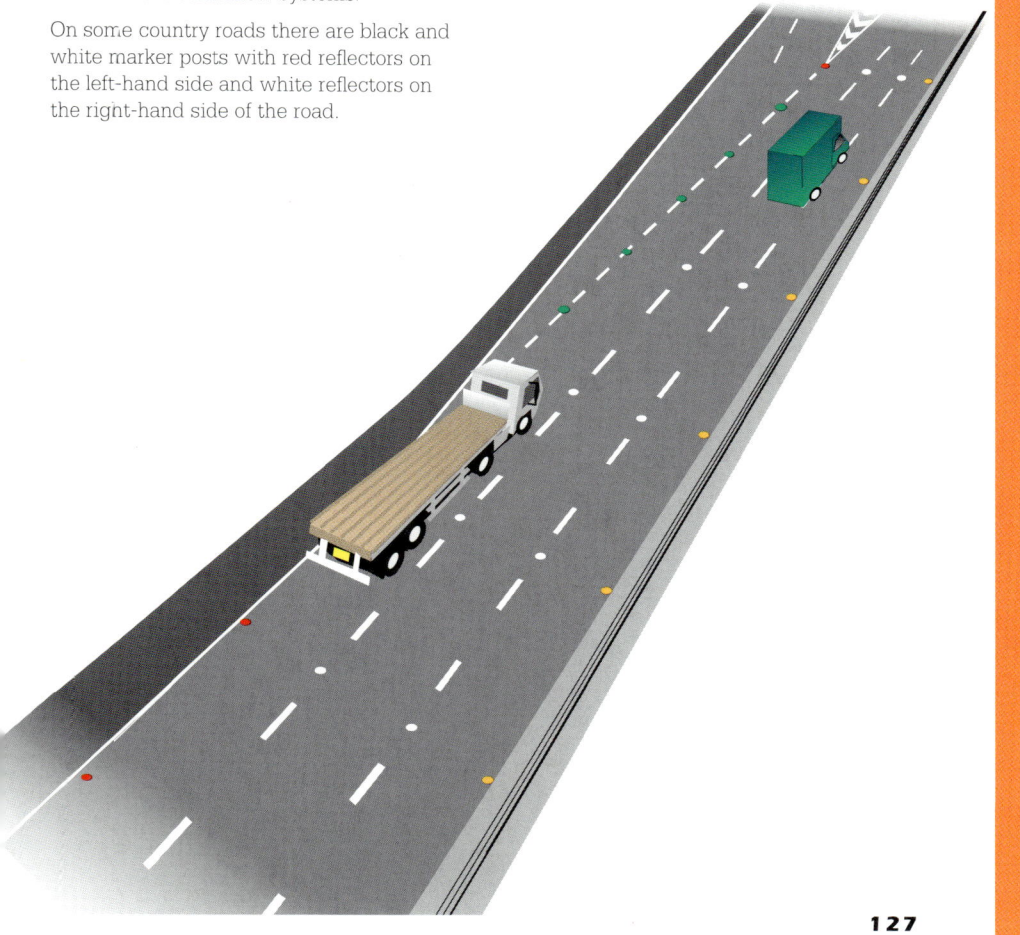

127

DRIVING SKILLS

High winds

In bad weather it's a good idea to listen to, watch or read the weather forecast if you're going to drive

- a double-decker bus or coach
- a high-floor coach
- a light or empty bus, coach or minibus.

If you have to drive on roads that often have strong winds such as

- high bridges
- high-level roads
- exposed viaducts
- exposed stretches of motorway

listen to advance weather warnings. Ferry crossings will also be affected by very strong winds. There could be delays or cancellations, so it's a good idea to check before setting off.

Watch out for signs warning of high winds, and beware of fallen trees or damaged branches that could fall on your vehicle. Take notice of signs and warnings and remember that

- roads may be closed to certain large vehicles
- there may be delays due to lanes being closed. This is done on high bridges to create empty 'buffer' lanes in the event of any large vehicles being blown off course
- you may need to use another route.

If you ignore any signs or warnings you could put your passengers, your vehicle and yourself at risk. If there's an accident your passengers could be injured and you could be prosecuted and convicted.

Don't take risks.

Other road users

When it's very windy other road users are likely to be affected when

- they overtake you
- you overtake them.

Check your mirrors as you overtake to see that they still have control of their vehicle. Also, watch out for vehicles or motorcycles 'wandering' into your lane.

Don't ignore warnings of severe winds.

You can't afford to take risks.

DRIVING SKILLS

Breakdowns

If your vehicle breaks down try to stop as far to the left as possible. If you can, get off the main carriageway without causing danger or inconvenience to other road users, especially pedestrians.

Move your passengers as far forward in the vehicle as you can. This should help to limit injuries if another vehicle runs into the back of yours.

Place a warning cone, pyramid or reflective triangle at least 45 metres (147 feet) behind the vehicle on normal roads, but do not attempt to place any warning device on a motorway carriageway, hard shoulder or slip road.

Some foreign-built buses and coaches have outside fuse and relay boxes on the right-hand side of the vehicle. Don't attempt to work on the right-hand side of the vehicle unless protected by a recovery vehicle with flashing amber beacons. Even then, take great care on roads carrying fast-moving traffic. Many accidents happen at breakdowns. Protect yourself, your passengers and your vehicle.

DRIVING SKILLS

Assessing the dangers

If your vehicle is creating an obstruction or is a potential danger to other road users tell the police as soon as possible. This is particularly important if your vehicle is carrying passengers, especially school children. Their safety must come first.

If you think that there's a serious risk of collision, escort your passengers off the bus. Ensure that they wait somewhere well away from the traffic. Explain carefully what you're doing and ask people to go for help if necessary.

Make sure that you

- know where all your passengers are
- know what they're doing
- keep them informed.

Don't leave them, unless absolutely necessary.

Recovery agencies

If you're driving long distances or on overnight services you must know what to do if you break down and require

- a replacement vehicle for your passengers
- the attendance of a breakdown vehicle and/or recovery.

If you're an operator, even if you have only one vehicle, you must be prepared for anything that might happen. Under no circumstances must passengers be left stranded.

Vehicles that break down on the motorway must be removed promptly for safety reasons.

Don't ignore danger signals

If you suspect that there's something wrong with your vehicle don't be tempted to carry on driving. You could end up causing traffic jams if your bus eventually breaks down in an awkward place.

A minor problem could turn out to have major effects. For example, a broken injector pipe dripping fuel onto a hot exhaust manifold may only seem to be a slight engine hesitation to the driver. However, this problem has been known to cause fires in which the vehicle was completely destroyed.

DRIVING SKILLS

Blow-outs

Many PCV breakdowns involve a tyre bursting commonly known as a 'blow-out'. These are dangerous because they make a bus difficult to control. They also leave debris on the road, which causes danger to other road users.

Front wheel blow-outs

A front wheel blow-out can mean that you won't be able to steer the bus properly. If this should happen you should

- keep a tight hold on the steering wheel
- always be aware of anything on the left-hand side of your bus
- signal left
- try to steer to the left-hand side of the road (or to the hard shoulder on the motorway)
- slow down gradually – don't brake hard
- try to stop your bus under control as far to the left as you can
- (if you need to) put a warning triangle, cones, or other permitted device behind the vehicle
- switch on the hazard warning lights if your vehicle is blocking part of the road.

If you can avoid braking hard or swerving you should be able to stop the bus without skidding.

Rear wheel blow-outs

If a rear tyre bursts you might not notice that it's happened. This is because most large vehicles have twin rear wheels. If you carry on driving, the second tyre on that side of the axle could also burst, as it's not designed to run on its own.

Although a rear wheel blow-out usually has less effect on the steering than a front wheel blow-out, the ride will become bumpy. Always try to find out what's causing odd handling.

Follow the same procedure for a front tyre blow-out, and pull off the road as safely as possible.

Safety checks

It's essential to make sure that all wheel nuts are tightened with the approved calibrated torque wrench. The wheel nuts should be checked every day before starting your journey.

Further information is given in the British Standard Code of Practice for the selection and care of tyres and wheels for commercial vehicles. This has been developed with the support and involvement of the major transport operators' associations. The relevant reference number is BS AU 50: Part 2: Section 7a: 1995, and is available from

British Standards Institution
389 Chiswick High Road
London
W4 4AL

Tel: 020 8996 9000
Website: www.bsi-global.com

DRIVING SKILLS

Stay alert and try to anticipate the actions of other road users. You need to understand how your vehicle will affect other road users, especially

- cyclists
- pedestrians
- motorcyclists.

Pedestrians standing on the edge of a kerb and cyclists are more vulnerable to being hit by your mirrors or being drawn under your wheels.

Assess every risk and try to eliminate it. You can remove most of the accident risk from your own driving by

- concentrating
- driving safely and sensibly
- staying alert
- being fully fit
- planning well ahead
- observing the changes in traffic conditions
- driving at a safe speed to suit the road, traffic and weather conditions
- keeping your vehicle in good overall condition
- making sure that passengers don't distract you
- not rushing
- avoiding the need to act hurriedly.

If you're involved in an accident

you must stop.

It's an offence not to do so.

Accidents

Drive at all times with anticipation and awareness. By driving defensively you lessen the risk of an accident.

If, however, you're involved in or have to stop at an accident, you should act decisively and with care to prevent any further damage or injury. Ultimately, your own safety and that of others must be your first concern.

DRIVING SKILLS

At an accident scene

If you're one of the first to arrive at an accident scene your actions could be vital. Find a safe place to stop, so that you do not endanger yourself, your passengers or other road users. You must ensure that either you or others

- warn other traffic by using hazard warning lights, beacons, cones, advance warning triangles, etc.
- check that there are no naked lights, or take the correct action if there are
- telephone 999, giving full details of what has happened
- check that all hazard flashers can be seen. If other road users confuse your signals it could make things worse
- switch off all engines
- stop anyone from smoking.

Dealing with injuries

It's best to avoid moving injured people until the emergency services arrive. You should be extremely careful about moving casualties – it could prove fatal. Casualties should only normally be moved if

- they're in need of resuscitation (that is, not breathing)
- in immediate danger (from fire, chemicals, fuel spillage, etc.).

You should

- move any apparently uninjured people away from the vehicle(s) to a safe place
- give first aid if anyone is unconscious (see pages 138–140)
- check for the effects of shock. A person may appear to be uninjured but might be suffering from shock
- keep casualties warm but don't give them anything to eat or drink
- give the facts (not assumptions, etc.) to medical staff when they arrive.

You should not remove a motorcyclist's helmet unless it is essential to do so.

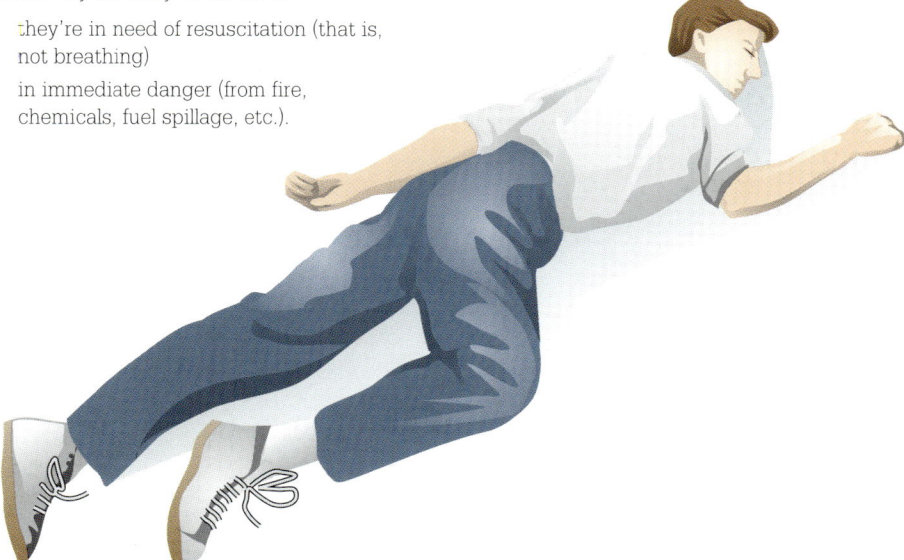

133

DRIVING SKILLS

Caring for passengers

You must do everything you can to protect your passengers at a breakdown or scene of an accident. Decide if there's any further danger and how best to reduce the risk. Tell passengers what's happening

- without upsetting them further
- by only giving them accurate information that they need to know.

You'll need to decide whether it's appropriate for passengers to

- stay where they are
- move to a safer position in the bus, if they're able (e.g., to the front if another vehicle could run into the back)
- get off the bus carefully and wait in a safe place, which you must select.

If you're unable to supervise the movement of your passengers, ask someone responsible to do it for you. You must not allow people to wander around. They could put themselves at risk or get in the way of the emergency services.

You should ask for people with medical qualifications to come forward and help.

On the motorway

Because of the higher speeds on motorways there's more danger of an accident turning into a serious incident. You must inform the motorway police and emergency services as quickly as you can.

- Use the nearest emergency telephone, this is connected directly to the police.
- If you use a mobile phone, identify your location from marker posts on the hard shoulder first.
- Don't cross the carriageway to get to an emergency telephone.
- Try to warn oncoming traffic, but don't endanger yourself.
- Move any uninjured people well away from the main carriageway and onto an embankment, etc.
- Watch out for emergency vehicles coming along the hard shoulder.

Emergency vehicles

Be aware that emergency vehicles may approach at any time while you are on the road. You should look and listen for flashing blue, red or green lights, headlights or sirens being used by ambulances, fire engines, police or other emergency vehicles. When one approaches do not panic; consider the route it is taking and take appropriate action to let it pass. If necessary, pull to the side of the road and stop, but make sure you are aware of other road users and that you do not endanger them in any way.

If you see or hear emergency vehicles in the distance, be aware that there may be an accident ahead and that other emergency vehicles may be approaching.

DRIVING SKILLS

Dangerous goods

If an accident involves a vehicle displaying either a hazard warning information plate or a plain orange rectangle

- give the emergency services as much information as possible about the labels and any other markings
- contact the emergency telephone number on the plate of a vehicle involved in any spillage, if one is given
- keep well away from such a vehicle. In attempting to rescue a casualty you may become one
- beware of any liquids, dusts or vapours – no matter how small the amount may appear to be. People have been seriously injured from just a fine spray of corrosive fluid leaking from a pinhole puncture in a tanker
- don't use a mobile phone close to a vehicle carrying flammable loads.

Documents and information

If your vehicle is involved in an accident you must stop. It's against the law not to do so. Also, you must

- inform the police as soon as possible, or in any case within 24 hours, if
  - anybody is injured
  - damage is caused to another vehicle or property and the owner isn't present or can't be found
  - the accident involves any of the animals specified in law
- produce your insurance documents and driving licence, and give your name and address to any police officer who may require it
- give these details to any other road user involved in the accident if they have grounds to ask for them.

If you can't show your documents at the time, whether anyone is injured or not, report the accident to the police as soon as you can, or in any case within 24 hours (in Northern Ireland you must report the accident to the police immediately).

You must inform the police as soon as possible and in any case within 24 hours (you must do this immediately in Northern Ireland), if

- there's injury to any person not in your vehicle
- damage is caused to another vehicle or property and the owner is either not present or can't be found easily
- the accident involves any of the animals specified by law.

The police may ask you to take your documents to a police station of your choice within seven days (five days in Northern Ireland), or as soon as is reasonably possible if you're already on a journey that takes you out of the country.

At the accident scene you must

- exchange details with any other driver or road user involved in the accident
- obtain names and addresses of any witnesses who saw the accident.

135

DRIVING SKILLS

Take notes at the scene so that you have the information when you need it. Make a note of

- the time
- the place
- street names
- vehicle registration numbers
- weather conditions
- lighting (if applicable)
- any road signs or road markings
- road conditions
- damage to vehicles or property (see page 61 for the procedure for railway bridge collisions)
- traffic lights (colour at the time)
- any indicator signals or warning (horn)
- any statements made by other people involved
- any skid marks, debris, etc.

Fire

Fire can occur on PCVs in a number of locations

- engine
- passenger areas
- kitchens and serveries
- toilets
- crew sleeping accommodation
- luggage lockers
- transmission
- tyres
- fuel system
- electrical circuits.

It's vital that any outbreak is tackled without delay. A vehicle can be destroyed by fire within an alarmingly short period of time.

If a fire is suspected or discovered, in order to avoid danger to others, it's essential to

- stop as quickly and safely as possible
- get everyone off the PCV as quickly as possible. Tell them to stand in a safe place
- either telephone 999 or get someone else to do it immediately
- tackle the source with a suitable fire extinguisher, **if you can do so safely. Do not endanger your own life**.

If the fire involves a vehicle carrying dangerous goods

- the driver must have been given training to deal with such an emergency. Follow their advice
- the vehicle should carry special fire extinguishers
- keep the public and other traffic well away from the fire
- isolate the vehicle, if you can, to reduce danger to the surrounding area
- make sure that someone calls, immediately, the emergency telephone number given on the hazard warning plate or the load documents
- warn oncoming traffic.

Stay calm – act quickly.

136

DRIVING SKILLS

Old-style UK fire extinguishers

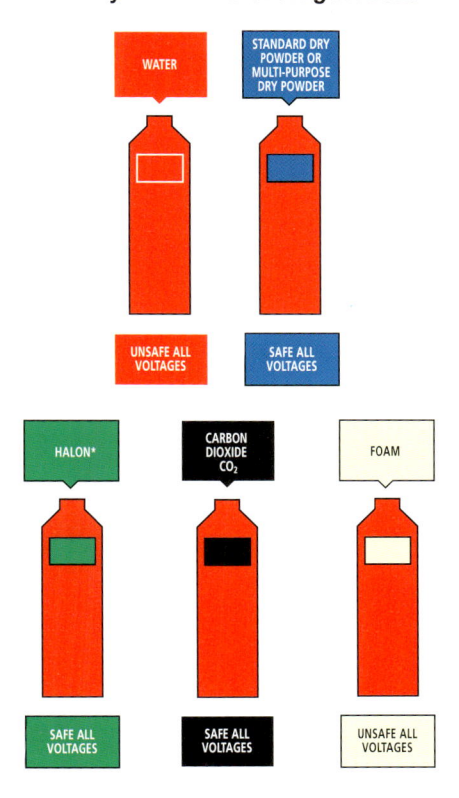

Vehicles carrying dangerous goods and other materials which may pose a hazard are subject to detailed emergency procedures which must be followed. Never put yourself in danger when tackling a fire. Always call the fire sevice as quickly as possible because they are the experts. Make sure any passengers leave the vehicle and go to a place of safety.

*Note

Halon fire extinguishers may still be used. However, halon is no longer manufactured in the EU for environmental reasons. Once used, a halon extinguisher cannot be refilled and should be replaced with a suitable alternative, such as a dry powder extinguisher.

Fire extinguishers

All PCVs must have at least one fire extinguisher. You must know where they're located and how to get them out and use them.

Regulations specify the type and size of fire extinguisher that must be carried on a bus or coach. You should be able to recognise the various types of fire extinguisher and know which fires they're intended to tackle. For example, it's dangerous to tackle a fuel fire with a water or carbon dioxide fire extinguisher, since this may only spread the fire further.

Most extinguishers are intended to smother the source of the fire by either the action of an inert gas or a dry powder. Try to isolate the source of the fire. If at all possible

- disconnect electrical leads
- cut off the fuel supply.

Don't open an engine housing wide if you can direct the extinguisher through a small cap. Also, avoid operating a fire extinguisher in a confined space.

New-style UK fire extinguishers

137

DRIVING SKILLS

First aid

Buses and coaches should carry first aid equipment, but it is not a legal requirement on a vehicle being used to operate a local service. You must know

- where it is
- how to get at it (if it's kept behind glass or in a safety compartment)
- what's in it
- how and when to use it.

As a professional driver, you're encouraged to take some first aid training. It could help save a life. There are courses available from the

- St John Ambulance Association and Brigade
- St Andrew's Ambulance Association
- British Red Cross Society.

The following information may be of general assistance, but there's no substitute for proper training.

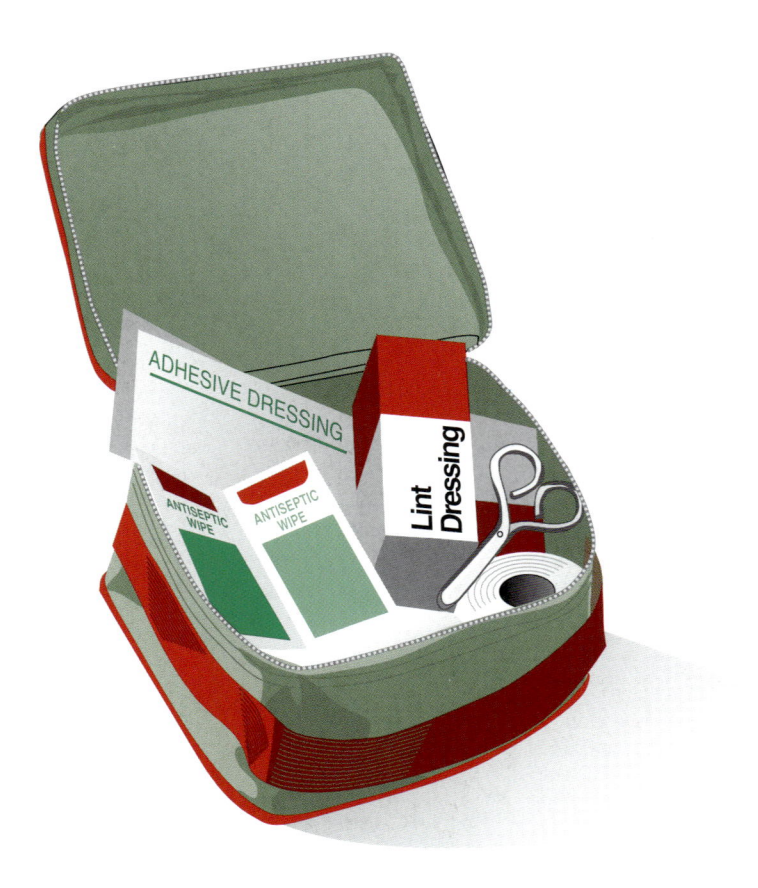

138

DRIVING SKILLS

First aid on the road

Any first aid given at the scene of an accident should only be looked on as a temporary measure until the emergency services arrive.

If you haven't any first aid training the following points could be helpful.

Accident victims

It is essential that the following are given immediate priority if the casualty is unconscious and permanent injury is to be avoided.

Remember the letters A B C

- **A** – the airway must be cleared and kept open
- **B** – breathing must be established and maintained
- **C** – circulation must be maintained and severe bleeding stopped

Airway Check for and relieve any obstruction to breathing. Unless you suspect head or neck injury, remove any obvious obstruction in the mouth (false teeth, chewing gum, etc.)

Breathing Breathing should begin and colour improve. If there's no improvement after the airway has been cleared

- tilt the head back very gently
- place a clean piece of material, such as a handkerchief, over the injured person's mouth
- pinch the casualty's nostrils together
- blow into the mouth until the chest rises. Take your mouth away and wait for the chest to fall
- repeat regularly once every four seconds until the casualty can breathe without help

With babies and small children

- let your mouth surround their mouth and nose and breathe very gently
- take your mouth away and wait for the chest to fall
- withdraw, then repeat regularly once every four seconds until breathing restarts and the casualty can breathe without help

Circulation Prevent blood loss to maintain circulation. If bleeding is present follow the procedure on page 140 to stem it.

Unconscious and breathing

Do not move a casualty unless there's further danger. Movement could add to spinal/neck injury.

If breathing is difficult or stops, treat as recommended in the breathing section.

Don't attempt to remove a motorcyclist's safety helmet unless it's essential – otherwise serious injury could result.

139

DRIVING SKILLS

Bleeding

To stem the flow of blood apply firm pressure to the wound without pressing on anything that may be caught in or sticking out from the wound.

As soon as practical fasten a pad to the wound with a bandage or length of cloth. Use the cleanest material available.

If a limb is bleeding, but not broken, raise it to reduce the flow of blood. Any restriction of blood circulation for more than a short time could cause long-term injuries.

It is vital to obtain skilled medical help as soon as possible. Make sure that someone dials 999.

Dealing with shock

The effects of shock may not be immediately obvious. Warning signs to look for include

- rapid pulse
- pale grey skin
- sweating
- rapid shallow breathing.

Prompt treatment can help to deal with shock

- don't give the casualty anything to eat or drink
- reassure the victim confidently and keep checking on them
- keep casualties warm and make them as comfortable as you can
- talk firmly and quietly to anyone who's hysterical
- don't let shock victims wander into the path of other traffic
- try not to leave any casualty alone
- don't move the casualty unless it's necessary
- if a casualty does need to be moved for their own safety, take care to avoid making their injuries worse.

Burns

Check the casualty for shock, and if possible, try to cool the burn. Try to find a liquid that is clean, cold and non-toxic with which to douse it.

Do not try to remove anything which is sticking to the burn.

DRIVING SKILLS

Electric shock

Some accidents involve a vehicle hitting overhead cables or electrical supplies to traffic bollards, traffic lights or street lights. Make a quick check before trying to get someone out of a vehicle in such cases.

Don't touch any person who's obviously in contact with live electricity unless you can use some non-conducting item, such as a dry sweeping broom, etc. You must not try to give first aid until contact has been broken.

Part Five
Test preparation

The topics covered

- **Preparing for the driving test**
- **Applying for the test**
- **The official syllabus**
- **Revised legislation**

TEST PREPARATION

Preparing for the driving test

The standard required to pass the PCV driving test is high. After all, you'll be carrying passengers who will be relying on you to deliver them safely to their destinations. Also, the vehicles you'll be licensed to drive require extensive knowledge, skill and responsibility to be driven safely.

The PCV driving test has been carefully designed to assess whether you've reached the required standard. To be properly prepared for the driving test you should cover the specific aspects of the officially recommended syllabus (found on page 156) and combine this with practice on a wide variety of roads in different traffic conditions. You should be able to demonstrate to the examiner that you can deal with any situation that arises – safely, skilfully and without help or advice.

Training organisations

There are a number of training organisations concerned with passenger transport, which have established the highest standards of training for the PCV driver. In addition, several large operators have driver-training divisions.

If you wish to work for one of these operators and are accepted onto their scheme, you'll be trained to drive using company buses. You may have to pay for this training, or agree to work for the company for a certain time.

The company itself will arrange for your PCV driving test when your driving is good enough. Some operators even have examiners of their own who are authorised to conduct tests. Otherwise you'll be tested by a DSA examiner at a PCV driving test centre.

You can find details of a local training group by contacting

- the Confederation of Passenger Transport UK, whose address is at the back of this book
- your local bus or coach operators
- your local Learning and Skills Council
- advertisers in your local press or in trade directories.

You'll normally be offered an 'assessment drive', lasting an hour or two. The instructor will then suggest the length of course you'll need and the cost.

Contact more than one training organisation and compare schemes. Try to choose an instructor or organisation with an established reputation for the quality of their instruction and proven PCV test results. Ask what arrangements are made should you need additional training as a result of failing a PCV test. Also, talk to newly qualified PCV drivers about their training.

143

TEST PREPARATION

Training coverage

It's in your own interest to find out how comprehensive a training course will be before you enrol. The opportunity to drive a variety of vehicles will obviously widen your knowledge and understanding of buses and coaches. In addition, your training should cover driving

- on as many different types of road as possible, including motorways
- in all sorts of driving conditions, including darkness
- on dual carriageways, where the upper speed limit for PCVs applies. (You'll probably be asked to drive on such roads during the PCV test.)

Whether you select operator training, a commercial driver-training school or an individual trainer, with perhaps only one vehicle, it's essential that all aspects of the syllabus set out in this part are covered. You should also have the opportunity to practise the braking and reversing exercises on a suitable off-road site. However, avoid concentrating solely on the off-road exercises.

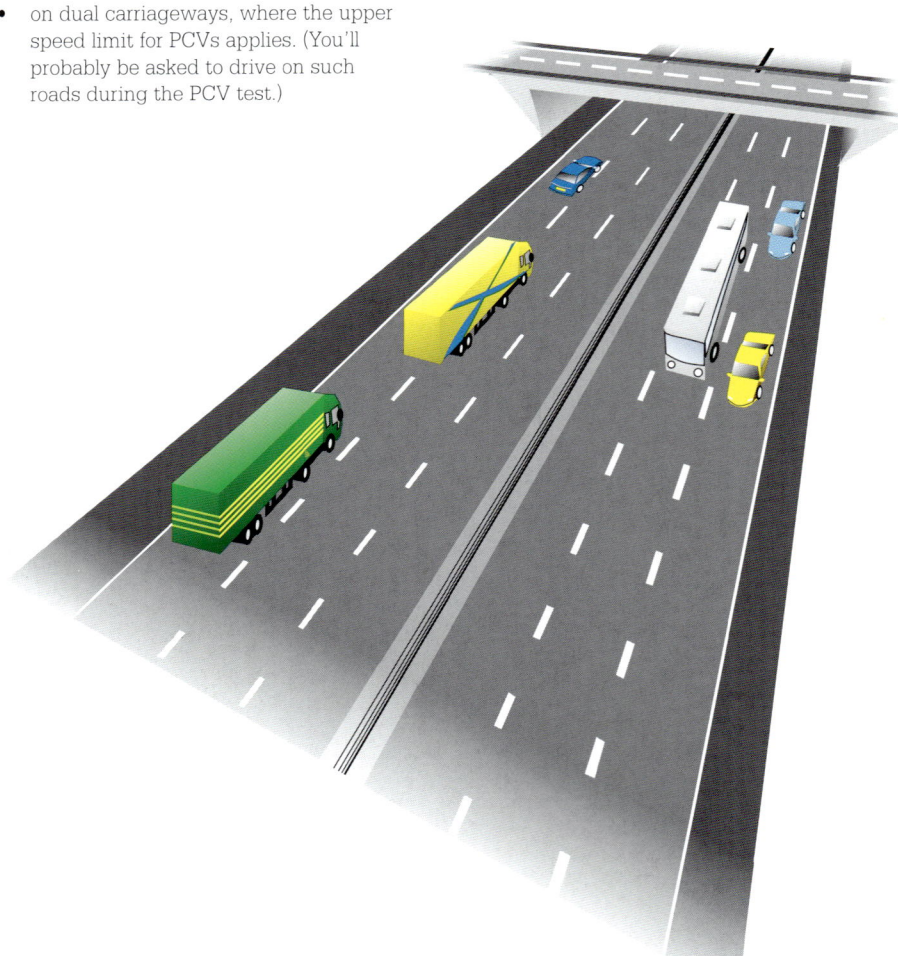

TEST PREPARATION

Causing a nuisance

Creating undue inconvenience for others should be avoided when you practise. Not all road users appreciate the difficulties that a bus driver faces when manoeuvring a large vehicle, especially

- moving off
- stopping
- turning left or right
- in narrow roads.

Causing a nuisance to residents and other traffic, or the continuous noise created by

- the hissing of air brakes or revving the engine to build up air pressure
- persistent stopping and starting

can soon become a reason for complaint in residential areas.

If a local problem exists due to PCV or LGV training already taking place, avoid making the situation worse. Your trainer should be aware of any such difficulties and use an alternative area to practise.

TEST PREPARATION

About the driving test

When taking the PCV practical test, you should aim for a professional standard. You'll pass if the examiner sees that you can

- drive safely and to a high standard
- show expert handling of all the controls
- carry out the set exercises accurately, under control and with good observation
- demonstrate a thorough understanding of *The Highway Code* and vehicle safety matters.

Does the standard of the test vary?

No. All examiners are trained to carry out tests to the same high standards nationally. Whether they're DSA or delegated company examiners, all are regularly checked to ensure that your driving will be assessed uniformly. You should have the same result whoever the examiner is and wherever the test takes place. In addition, test routes

- are as similar as possible
- include a wide range of typical road and traffic conditions.

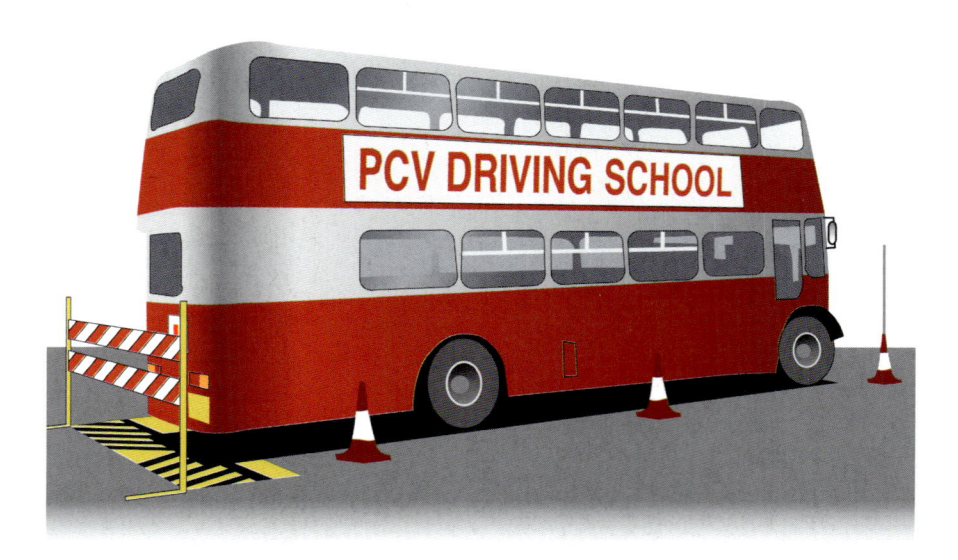

TEST PREPARATION

How your driving test is assessed

Your examiner will assess any errors you make. They will be assessed and recorded depending on their degree of seriousness and marked on the Driving Test Report form (DLV25).

You will fail your test if you commit a serious or dangerous fault. You will also fail if you accumulate too many driving faults (previously known as minor faults).

The criteria the examiner will use are as follows:

Driving fault – less serious but has been assessed as such because of circumstances at that particular time.

Serious fault – recorded when a potentially dangerous incident has occurred or a habitual driving fault indicates a serious weakness in a candidate's driving.

Dangerous fault – recorded when a fault is assessed as having caused actual danger during the test.

At the end of the test you will be offered some general guidance to explain your driving test report. Your instructor can be present during this debrief even if he or she has not accompanied you on the driving test.

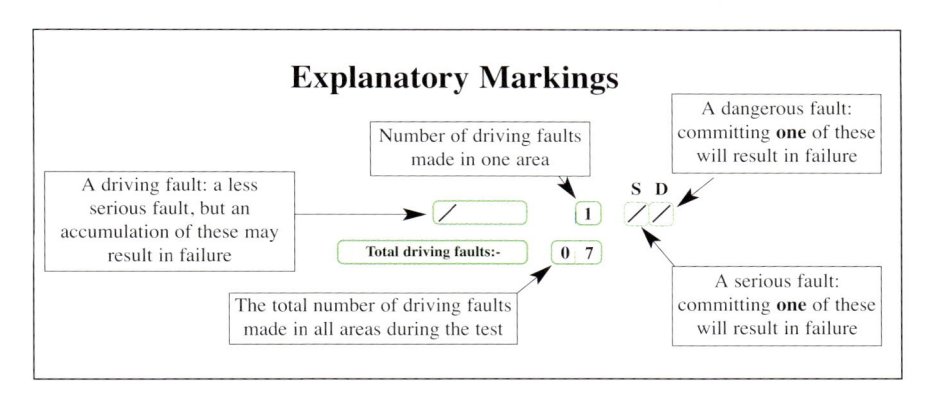

Explanatory Markings

A driving fault: a less serious fault, but an accumulation of these may result in failure

Number of driving faults made in one area

A dangerous fault: committing **one** of these will result in failure

A serious fault: committing **one** of these will result in failure

Total driving faults:- 0 7

The total number of driving faults made in all areas during the test

147

TEST PREPARATION

Are examiners supervised?
Yes, they are closely supervised. A senior officer may sit in on your test. Don't worry about this. The supervising officer won't be examining you, but will be checking that the examiner is carrying out the test properly. Just carry on as if he or she wasn't there.

Can anyone accompany me on the test?
Your instructor is allowed to accompany you but can't take any part in the test. Regulations prevent passengers other than those mentioned from being carried on a test.

What if I need an interpreter?
If you need an interpreter you should arrange for one to come with you. They must not be your driving instructor or anybody younger than 16 years of age. Time will be spent at the start of your test discussing the best way to give directions or instructions that are clearly understood.

How should I drive during the test?
Drive in the way your instructor has taught you. If you make a mistake, try not to let it worry you. It might be a trivial mistake and may not affect the results of the test. Your examiner will be looking for a high overall standard. Don't worry about one or two minor mistakes.

What will my examiner want from me?
Your examiner will want you to drive safely to a high standard under various road and traffic conditions. You'll be

- given directions clearly and in good time
- asked to carry out set exercises.

If at any time you are unable to hear or understand the instructions given, ask for them to be repeated, the examiner will not mind. He or she will try to put you at your ease.

What will the test consist of?
The test will last around 90 minutes. Apart from general driving, the test will include

- reversing within a marked area into a restricted opening
- a braking exercise
- a gear changing exercise, if you're driving a manual vehicle
- moving off on the level, at an angle, uphill and downhill
- demonstrating the uncoupling and recoupling procedure, if you're taking your test with a trailer.

From 1 September 2003 you will also need to satisfy the examiner that you're capable of preparing to drive safely by carrying out simple safety checks on the vehicle you're using on the test. The safety checks and some of the manoeuvres are carried out at the test centre. These are

- reversing within a marked area into a restricted opening
- a braking exercise
- an uncoupling or recoupling exercise if relevant.

The remainder of the exercises will take place during the road section of the test.

During the reversing exercise your examiner will remain outside the vehicle. Your examiner will join you in the cab before explaining the braking exercise to you. He or she will watch your handling of the controls and observations as you carry out the exercise. This exercise will be carried out before you leave the test centre. If your vehicle doesn't pull up satisfactorily your examiner may decide not to continue with the test, in the interest of safety.

If a delegated company examiner conducts your driving test, all the exercises may be carried out on public roads or at an agreed private site.

TEST PREPARATION

The Highway Code

You must know and understand *The Highway Code* thoroughly and put it into practice during the test. Study the latest edition carefully.

Questions on *The Highway Code* form part of the theory test for drivers of large vehicles. Training materials for the multiple choice part of the test also include *The Official Theory Test for Drivers of Large Vehicles* and *The Official Theory Test CD-Rom for Drivers of Large Vehicles*.

As well as the multiple-choice questions, the theory test now includes a hazard perception part. To prepare for this, DSA strongly recommends that you study and work through the hazard perception training material. There is a DVD available, entitled *The Official Guide to Hazard Perception* and, alongside this, there is a video and workbook-based training pack available, called *Roadsense*.

You will need to put into practice what you have learned for your theory test when taking the practical test.

You can start driver training before you apply for the theory test, but you must pass before you're permitted to apply for the practical PCV driving test. A theory test pass certificate is valid for two years.

149

TEST PREPARATION

Being fully prepared

Driver training for large vehicles is usually intensive, so it may be necessary for either you or your trainer to book your PCV driving test before you've reached the standard required to pass.

Your instructor may offer you a mock test shortly before your real test is due. This will give you an understanding of how the test will be conducted and may alert you to any weaknesses. Make sure that you understand what you're asked to do and, should you need to work on any problem areas, work with your instructor to overcome them.

Having a test date to aim for can be a good incentive. However, drivers acquire skills and understanding at differing rates and it's possible that you may need more time and training than you'd planned. If this happens, postpone your test; it is better to go in for your test feeling confident. Be advised by your trainer.

Do not leave it too late to cancel your test appointment, as a late cancellation may result in you losing your fee for the driving test.

Driving examiners are observers during a driving test – they are not there to advise you on how to drive, so make sure you feel confident in your own ability.

If your instructor doesn't feel that you have competent, safe control of the vehicle by the time of the test appointment, accept that judgement. You'll be advised about the options for additional training, and an alternative test appointment may be available to you.

TEST PREPARATION

Applying for the test

When you reach the standards set in this book – not before – apply for your practical test.

You should be driving

- consistently well
- with confidence
- in complete control
- without assistance and guidance from your instructor.

You will then be ready for your PCV driving test. Be advised by your instructor and make sure you have enough practice before you apply.

To apply for the test you must have an entitlement to drive PCVs (either a category D provisional licence or a full licence for a category that includes provisional entitlement for the category which you wish to be tested on). In addition, you'll need a valid theory test pass certificate.

Special circumstances

If you're disabled in any way you'll still take the same PCV driving test as every other candidate. Your examiner may wish to talk to you about your disability and any adaptions fitted to your vehicle. For this reason it is important to give details of your disability when you apply for your test.

To make sure enough time is allowed for your test it would help DSA to know if

- you're restricted in any way in your movements
- you have a disability that may affect your driving.

If you would like further information please see the list of useful addresses at the back of this book.

Language difficulties

If you have difficulty speaking or understanding English you can bring an interpreter with you. The interpreter must be 16 years or over and must not be your instructor. Please include this information on your test application form.

TEST PREPARATION

The application form

You can obtain an application form (DLV26) for the driving test from any LGV/PCV test centre or by phoning the National Booking telephone number, 0870 01 01 372. Study the guidance notes (DL26/I) carefully, including the table of PCV categories, especially if you wish to drive vehicles in more than one category. It is important to note that if you pass the test in a semi- or fully automatic vehicle, you won't be able to drive vehicles with a manual gearbox.

Make sure that you give all the particulars required on the application form, otherwise it will only be returned to you. This will certainly delay your driving test appointment.

Send the correct fee with your application. Posters at test centres list the fees. Cheques or postal orders should be crossed and made payable to the Driving Standards Agency. If you send a postal order keep the counterfoil.

Don't send cash

Send the application form to

DSA
PO Box 280
Newcastle-Upon-Tyne
NE99 1FP

Make sure that you enclose the correct fee; your application may be delayed otherwise.

Send your application to DSA at least 28 days before your preferred date for the test (in summer, longer notice is often helpful due to the increased demand). Programmes of tests are arranged well in advance so, if you don't give enough notice, you may not be given your preferred date.

Visit the website

More information and guidance on all aspects of booking or taking a driving test can be found on the DSA website at **www.driving-tests.co.uk**

You can also book your theory test on the DSA website. An online practical test booking facility is currently under development.

TEST PREPARATION

Booking by credit or debit card

You can book your PCV driving test by telephone using a credit or debit card. Most major credit and debit cards are accepted. You must be the card holder; if you aren't, the card holder must be present.

For information about fees and to book an appointment, phone DSA's National Enquiries and Booking Centre on 0870 01 01 372.

You'll find it helpful to have filled in an application form (DLV26) before making the call. You'll definitely be asked for the information listed on the form. Your appointment date and time will be given to you over the telephone. You should receive written notification confirming the appointment within a few days.

Trainer booking

DSA has a facility for trainers to book tests for candidates. Ask at your training school whether you're able to take advantage of the scheme or if you should book the test for yourself.

If you take the test with a delegated examiner from an operator's premises, the test will usually be offered to you at the end of the course. The cost will normally be included as part of your agreement with the company, but you may be asked to pay a separate fee.

Trainers wishing to use the trainer booking system should contact DSA's Booking Section by telephone on 0870 01 01 372.

TEST PREPARATION

Saturday and evening tests

Saturday and weekday evening tests are available at some PCV driving test centres. The fees for these are higher than for a test during normal working hours on weekdays. You can get details from

- the National Booking telephone number 0870 01 01 372
- PCV driving test centres
- Your instructor.

Your test appointment

DSA will send notification of your appointment, which is the receipt for your fee. Take this with you when you attend your test.

It will include

- the time and place of your test
- the address of the driving test centre
- other important information.

If you do not receive notification after 21 days, contact the National Booking telephone number 0870 01 01 372.

Check your appointment notification as soon as you receive it to make sure that the date and time of the test appointment are suitable. If you need to postpone or change the appointment, you should notify DSA as soon as possible and return your appointment notification.

To cancel, you must give at least ten clear working days' notice. That means ten whole working days – not counting the day the DSA received your notification nor the day of your test. If you don't give enough notice you'll forfeit your fee and will have to re-apply with another fee.

Change of address or vehicle

Please telephone the National Booking telephone number immediately if you change your address before the day of your appointment. Also, you'll need to inform DSA if you have to bring a different vehicle from the one described on your application form. Otherwise, there could be a delay when you arrive for your test.

Inform the driving examiner at the test centre, either beforehand or as soon as you arrive, if there's any last-minute change of vehicle.

TEST PREPARATION

Extended tests

If you're found guilty of certain driving offences the courts may direct you to retake your PCV driving test. For some offences, which involve disqualification from driving for a period of time, you may need to take an extended car driving test. This means that

- it will be necessary to apply to the DVLA for a provisional licence entitlement
- you may only apply for a PCV test after passing an extended category B driving test

- you'll have to pass a normal PCV driving test if you previously held a PCV entitlement and wish to regain it.

There are higher fees for extended car driving tests, but not for the associated PCV driving test.

Remember, if you lose your category B (car) entitlement you'll lose your PCV entitlement. Your PCV entitlement may be returned on passing the category B test, but this is at the discretion of the Secretary of State for Transport.

TEST PREPARATION

The official syllabus

This syllabus lists the skills and knowledge required to be a good bus or coach driver and to pass the PCV practical driving test. Use the syllabus as a check-list while training.

Make sure that you understand all the areas covered. Other parts of this book explain in more detail the various topics in the syllabus. Your instructor will be able to answer any queries you have regarding preparation for your test.

During your driving test you won't be tested on all the items listed in the syllabus. However, you do need to understand them all. You need to know about **all** the aspects of being a safe and professional driver. While you're driving your examiner will want to see that you put your knowledge into practice. Think of passing the test as only one stage in becoming a good driver.

If you drive a PCV for which no special driving test is needed, this syllabus will help you to achieve the high standard of driving required for your own safety and that of your passengers.

Knowledge

You must have a thorough knowledge and understanding of

1. The latest edition of *The Highway Code*.
2. Regulations governing drivers' permitted hours (EC 3820/1985).
3. Regulations relating to the carriage of passengers (Public Passenger Vehicles Act 1981 and 1990 amendments).*
4. General motoring regulations, especially
   - road traffic offences
   - producing driving licences
   - holding operator's and road fund licences, and displaying discs where applicable
   - holding and displaying community bus permits, where applicable
   - insurance requirements (including 'green cards' or bail bonds that may be needed when abroad)
   - the Temp 100 regulations, if you intend to drive outside the UK
   - the information required to be shown on PCV manufacturers' plates
   - annual testing and the certification requirements for tachographs and speed limiters
   - the importance of regular vehicle maintenance and defect reporting procedures.
5. Health and Safety legislation, as it applies to PCV duties.

*Note
Certain minibuses, mobile project buses, playbuses, historic vehicles and community buses are subject to a relaxation of the Public Service and Passenger Carrying Vehicle regulations. If you drive one of these vehicles you must be aware of any restrictions on its use.

If your vehicle is equipped with a trailer, you must know which legal requirements apply.

156

TEST PREPARATION

You must also have a basic understanding of the function of the component parts of a PCV, including

6. Internal combustion engines
   - petrol
   - diesel
   - other fuels.
7. Power and control units in electrically propelled vehicles, if appropriate.
8. Ancillary and control systems.
9. The body and its equipment.

Legal requirements

To learn to drive a PCV you must

1. Be at least 21 years old.*
2. Meet the stringent eyesight requirements.
3. Be medically fit to drive PCVs of any type.
4. Hold a full car licence (category B or, if issued prior to 1990, group A).
5. Hold and comply with the conditions for holding either
   - a provisional PCV category D entitlement, or
   - a full PCV entitlement for another category of vehicle, which confers provisional entitlement for the vehicle you wish to drive.
6. Be sure that any vehicle driven
   - is legally roadworthy
   - has the required manufacturer's plate
   - has a current test certificate that covers its use
   - is properly licensed and has the correct tax disc displayed (and 'O' licence or permit disc, if required)
   - complies with the requirements of the tachograph and speed limiter legislation and displays the required certificates, if applicable
   - meets minimum vehicle requirements if used for a PCV driving test (see page 240).
7. Make sure that the vehicle being driven is properly insured for its use, especially if it's on contract hire.

*Note

You may learn to drive a PCV and take the driving test between the ages of 18 and 21, but if you pass you aren't permitted to carry passengers unless the vehicle is covered by a Public Service Vehicle operator's licence, a bus or community bus permit, and either

- the bus has no more than 16 passenger seats and you drive only in the UK, or
- the route mileage doesn't exceed 50 km (31 miles).

157

TEST PREPARATION

8. Display L plates to the front and rear of the vehicle (D plates, if you wish, when driving in Wales).
9. Be accompanied by a supervisor who holds a valid full UK licence for the category of vehicle being driven.
10. Be aware of the legal requirement to notify the DVLA of any medical condition that could affect safe driving.
11. Ensure that all information required on the vehicle by law (referred to as the 'legal lettering') is displayed, as applicable
    - seating/standing capacity
    - emergency exit location
    - fuel cut-off switch
    - electrical isolator switch
    - first aid equipment
    - fire extinguisher(s)
    - unladen weight of vehicle
    - height, displayed in the cab if the vehicle is over 3.0 metres (9 feet 10 inches)
    - registered company name and address
    - engine stop button.

You must also avoid

12. Using any mobile telephone or radio transmitter whilst driving the vehicle (except for limited use of Band III radio systems used for route control and emergency purposes).
13. Stopping on the hard shoulder of a motorway to use any mobile telephone or radio transmitter (unless in an emergency).
14. Using any public address system fitted in the vehicle to give any commentary whilst driving (except for brief location information which may be given using a 'hands-off' system).
15. Driving the vehicle whilst
    - issuing tickets
    - giving change
    - holding a conversation, other than in an emergency
    - being distracted
    - smoking
    - passenger doors are open.
16. In addition, you must know and apply the legal requirements relating to the vehicle and its use, where applicable, in respect of
    - speed limits
    - seating/standing capacity
    - fire extinguishers
    - first aid equipment (location and use)
    - interior lighting during the hours of darkness
    - the carriage and consumption of alcoholic drinks
    - the emptying of toilet waste storage tanks
    - hazardous substances that may be brought on board by passengers.

TEST PREPARATION

Vehicle controls, equipment and components

You must

1. Understand the function and use of the main controls of the vehicle
   - accelerator
   - clutch, if applicable
   - gears
   - footbrake
   - handbrake
   - steering, including power-assisted steering

   and be able to use them competently.

2. Know the effects speed limiters will have on the control of your vehicle, especially when you intend to overtake.

3. Know the principles of the various systems of retarders that may be fitted to PCVs
   - electric
   - engine-driven
   - exhaust brakes

   and when they should be brought into operation.

4. Know the function of all other controls and switches on the vehicle and be able to use them competently.

5. Understand the information given by
   - gauges
   - warning lights or buzzers
   - other displays on the instrument panel.

6. Be familiar with the operation of tachographs and their charts and any other time, speed or distance recording equipment that may be fitted. You should know what action to take if a fault develops in this equipment.

7. Know which checks should be made before starting a journey.

8. Know the safety factors relating to
   - seated and standing passengers
   - loading
   - stability
   - controls of any driver-operated doors
   - stowing luggage when passengers are carried.

9. Be able to carry out routine safety checks and identify defects, especially with the
   - engine performance
   - fuel systems
   - lubricating systems and oil levels
   - coolant temperature and levels
   - exhaust systems
   - gearbox operation, controls and transmission
   - braking system efficiency and operation
   - steering (including power-assisted systems)
   - suspension
   - tyres, wheel security and mudguards
   - heating, air conditioning and ventilation
   - air tanks (air pressure)
   - electrical systems, including
   - lights
   - direction indicators
   - destination displays
   - wipers and washers
   - bells, buzzers and linked 'bus stopping' displays
   - 'emergency exit insecure' warning devices, if fitted
   - horns
   - fuses, cut-outs and relays

159

TEST PREPARATION

- windscreen
- reflectors
- exterior bodywork, panels, fittings and trim
- service doors
- side and rear-view mirrors
- interior bodywork, seating, fittings and trim
- floor coverings
- emergency exits
- first aid equipment
- fire extinguisher
- vehicle loading

and, where fitted,

- seat belts and grab rails
- equipment for wheelchair access and security
- mechanically, electrically or air-operated doors
- adjustable suspension on 'kneeling' vehicles
- securing devices on emergency doors
- equipment for breaking emergency windows
- staircases.

Road user behaviour

You must know how to limit the risk of being involved in a road traffic accident by understanding

1. The most common causes of those accidents
2. Which road users are more vulnerable, for example
   - children
   - young riders and drivers
   - elderly drivers
   - elderly or infirm pedestrians
   - cyclists and motorcyclists
   - learner drivers.
3. The rules, risks and effects of drinking before driving.
4. The effects on your performance of
   - illnesses
   - drugs
   - cold remedies
   - other medication
   - tiredness.
5. The importance of complying with rest period regulations.
6. How to
   - concentrate
   - plan ahead
   - anticipate the actions of other road users.

TEST PREPARATION

Vehicle characteristics

You must know

1. The most important principles concerning braking distances under various road, weather and loading conditions.
2. The different handling characteristics of vehicles with regard to
   - speed
   - stability
   - braking
   - manoeuvrability
   - turning circles.
3. That some other vehicles, such as cycles and motorcycles, are less easily seen than others.

You must also be aware of

4. Blind spots that occur on many large vehicles.
5. The need to be extra vigilant when reversing any PCV into or out of a bay at boarding points or in workshops.
6. The safe angle of tilt, which must not be exceeded when driving high vehicles.
7. The risks and difficulties presented when
   - long vehicles negotiate speed reduction humps or humpback bridges
   - high vehicles are driven along roads with an adverse camber, thus leading to possible collisions with
   - shop blinds
   - buildings
   - road signs
   - traffic lights
   - telephone poles
   - overhead cables
   - trees
   - lamp standards
   - scaffolding
   - other high vehicles
   - vehicles with large mirrors pass close to pedestrians, street 'furniture' or other vehicles
   - heavy vehicles drive on, or close to, soft or damaged verges
   - the vehicle being driven encounters the minimum clearance needed under bridges.
8. The difficulties caused by the characteristics of both your own and other vehicles, and be able to take the appropriate action to reduce any risks that may arise.

Examples of situations requiring special care are when
   - long wheel-base coaches, buses and large goods vehicles move to the right before making a sharp left turn
   - articulated vehicles take an unusual line before negotiating corners, roundabouts or entrances
   - short wheel-base vehicles with front and rear overhang turn left or right, or when at bus stops, lay-bys, pedestrian crossings, etc.
   - cycles, motorcycles and high-sided vehicles are buffeted in strong winds, especially on exposed sections of road
   - turbulence created by coaches, double-decker buses and large goods vehicles travelling at speed affects
   - pedestrians
   - cyclists
   - motorcyclists
   - vehicles towing caravans
   - smaller vehicles.

161

TEST PREPARATION

Road and weather conditions

You must

1. Know about the hazards that can arise when driving on various types of road with differing volumes of traffic, such as
   - country lanes
   - single-track roads
   - one-way streets
   - those with bus lanes
   - contraflow systems
   - those in built-up areas
   - three-lane roads
   - dual carriageways with various speed limits
   - trunk roads with two-way traffic
   - motorways
   - roads or reserved areas where light rapid transit vehicles (LRTs or 'supertrams') operate
   - busways.
2. Know about the hazards that can arise when driving in various weather conditions, such as
   - strong sunlight
   - rain
   - snow and ice
   - fog
   - wind, especially when driving high vehicles

   and at all times of the day and night.
3. Know which surfaces will provide better or poorer grip when accelerating and braking.
4. Drive sensibly and anticipate how the conditions may affect the driving of other road users.
5. Understand the need to be aware of other road users when pulling up at bus stops, especially near junctions.
6. Appreciate the need to give correct signals, especially before pulling up at
   - bus stops
   - road junctions
   - pedestrian crossings, etc.
7. Recognise the special risks when passengers board or alight from your vehicle, such as
   - schoolchildren
   - elderly people
   - the disabled
   - those with
   - babies
   - toddlers
   - pushchairs
   - luggage.
8. Be aware of the presence of other road users by making effective use of the mirrors and by looking round before moving off from a standstill. Watch out, in particular, for the passenger who attempts to board or alight as you move off.

TEST PREPARATION

Traffic signs, rules and regulations

You must

1. Have a thorough knowledge and understanding of the meanings of traffic signs and road markings, especially those relating to
   - bus lanes, which may also permit cycles and taxis
   - bus priority systems
   - light rapid transit systems.
2. Be able to recognise and comply with traffic signs that point out
   - weight limits
   - height limits
   - length limits
   - width limits
   - prohibited entry for motor vehicles
   - no left or right turns
   - loading/unloading restrictions
   - roads designated Red Routes
   - traffic calming measures.

Note
Some signs may exempt buses.

Vehicle control and road procedure

You must have the knowledge and skill to take the following precautions, some of which will require assistance.

1. Before getting into the vehicle check that
   - you have all the required paperwork (especially for foreign trips)
   - all required discs and certificates are displayed
   - there are no obstructions round your vehicle
   - the emergency exit(s) operate correctly and are closed securely
   - all bulbs, lenses and reflectors are fitted, clean and undamaged
   - all lights, including indicators and stop lights, are undamaged and working
   - tyres and wheel nuts are free from obvious defects (visual check)
   - all windows and mirrors are clean and free of traffic grime and cracks
   - all body panels are secure
   - all external lockers and crew compartment doors are secure
   - there are no fluid or air system leaks
   - fuel and electrical isolation switches are clearly marked and turned on
   - all route numbers and destination blinds or displays are correct (or replaced by information that indicates that the vehicle isn't in service).

Note
Controls for route and destination displays are usually on board the vehicle. Adjust them as necessary.

163

TEST PREPARATION

2. After entering the vehicle check
   - the correct operation of any warning device fitted to an emergency exit that isn't visible from the driving position
   - that the entrance and exit doors (if fitted) operate correctly, and that any warning systems work properly
   - the location of the fire extinguisher(s) and first aid equipment
   - that heating, air conditioning or ventilation equipment is working properly and set for the conditions
   - that the bell or buzzer signal and any passenger information system works
   - that all gangways and staircases are clean, clear and free from defects
   - that all seats are clean, secure and free from defects
   - that the interior lighting operates correctly, including the exit/entrance step lights
   - that equipment for wheelchair access is operational

   where these items are fitted, and
   - that any graffiti is removed at the earliest opportunity, especially if it might cause offence
   - that any luggage or equipment is safely stowed.

3. Before starting the engine check
   - that the handbrake is applied and the gear selector is in neutral or the 'start' position
   - your seat, if necessary, for
     - height
     - distance from the controls
     - support and comfort
     - maximum vision
   - the mirrors, if necessary, to give a clear view of
     - traffic behind
     - the entrance/exit
     - intending passengers
     - the upper deck, where appropriate
   - the doors (if fitted) are closed
   - seat belts (if fitted) are in use.

4. When you start the engine, but before moving off, check that
   - the vehicle lights are on, if required
   - gauges indicate correct pressures for braking and ancillary systems
   - no warning lights are showing, which indicate it's unsafe to drive the vehicle
   - no warning buzzer is operating
   - all fuel and temperature gauges are operating normally and that there's sufficient fuel for your journey
   - suspension systems are at the correct height, if appropriate
   - all doors are closed
   - all equipment operates correctly (wipers, washers, indicators, etc.)
   - special access facilities, such as kneeling suspension, ramps or lifts, are correctly adjusted or stowed
   - it's safe, by looking all round. Before moving off especially check
     - the blind spots
     - entry/exit door(s) or boarding platform(s)
     - near the wheels.

Note
Air-operated systems, such as suspension and doors, may come into operation as air pressure builds up. Ensure this happens safely.

164

TEST PREPARATION

5. At the first opportunity, and before carrying passengers, check the brakes and steering for correct and effective operation. Also check that exhaust emissions aren't excessive (when the engine's warm).
6. When driving you must be able to
   - move off safely
     - straight ahead
     - at an angle
     - on the level
     - uphill
     - downhill
   - select the correct road position and appropriate gear at all times
   - take effective observation in all traffic conditions and give appropriate signals, when necessary
   - drive at a speed appropriate to the road, traffic and weather conditions
   - anticipate changes in traffic conditions
   - take the correct action at all times and exercise care in the use of the controls
   - move into the appropriate traffic lane correctly and in good time
   - pass stationary vehicles safely
   - meet, overtake and cross the path of other vehicles safely
   - turn right or left, or drive ahead at junctions, crossroads or roundabouts
   - keep a safe separation gap when following other vehicles
   - act correctly at all types of pedestrian crossing
   - show proper regard for the safety of all other road users, particularly the most vulnerable
   - keep up with the flow of traffic where it's safe and appropriate to do so, whilst observing all speed limits
   - comply with
     - traffic regulations
     - traffic signs
     - signals given by authorised persons, police officers, traffic wardens or school crossing patrols
   - take the correct action on signals given by other road users
   - stop the vehicle safely at all times
   - show courtesy and consideration to passengers at all times, particularly those with special needs
   - wait until elderly or disabled passengers are seated
   - be aware at all times of the effects that harsh braking, acceleration or steering will have on passengers, especially those
     - standing
     - moving toward exits
     - moving away from entrances.
   - pay particular attention to the care of
     - the elderly
     - the disabled
     - mothers with babies or toddlers
   - cross safely all types of level crossings, such as railway or light rapid or railed transit systems (LRTs or 'supertrams')
   - select safe and suitable places to stop the vehicle close to the nearside kerb, as is practicable, when requested
     - on the level
     - facing uphill
     - facing downhill
     - before reaching a parked vehicle

165

TEST PREPARATION

- leave sufficient room to move away when the platform of the vehicle's close to passenger boarding points at bus stops and when requested on 'hail and ride' services
- stop the vehicle in an emergency
  - safely
  - as quickly as possible
  - under full control
  - within a reasonable distance
- reverse the vehicle
  - under control
  - with effective observation
  - accurately
- enter a restricted opening to the left or right, and stop with the extreme rear of the vehicle where required (when carrying out a reversing exercise with a delegated examiner)
- follow advertised timetables and, in particular, not depart early from published timing points.

7. You must be able to carry out, as necessary, all these checks and manoeuvres
   - safely and expertly
   - in daylight
   - during the hours of darkness.

   Where your actions may affect other road users you must
   - make proper use of the mirrors
   - take effective observation
   - give signals, when necessary
   - act predictably.

   For the PCV driving test you'll be asked to carry out specific exercises to demonstrate your ability to stop quickly and to reverse. If the test isn't conducted by a DSA examiner but by a delegated (company) examiner these manoeuvres may be carried out on the public roads.

8. Before leaving the driver's position you must make sure that
   - the vehicle is stopped in a safe, legal and secure place
   - the handbrake is on
   - the gear lever/selector is in neutral or 'park'
   - the engine is stopped
   - the keys have been removed from the starter switch, if applicable
   - the electrical system is switched off, unless lights or other systems are required (on some vehicles the switch may not be within reach of the driving position)
   - you won't endanger anyone when you open any door.

9. When leaving a vehicle make sure that
   - all windows are closed
   - the passenger door is secure (if fitted)
   - you take all possible precautions to prevent theft of the vehicle
   - any available anti-theft device is used (e.g., immobiliser/alarm)
   - you've selected a safe place to leave the unattended vehicle
   - the parking place is
   - legal (not a 'no waiting' zone)
   - safe (it won't cause any danger to others)
   - convenient (not blocking any access or exit)
   - suitable (level and firm enough to support the weight of the vehicle).

10. If you'll be leaving the vehicle but the public will still have access (for instance, on playbuses or mobile project vehicles) ensure that
    - the cab area is isolated
    - a responsible person is in attendance.

TEST PREPARATION

Additional knowledge

You must know

1. The importance of inspecting all tyres on the vehicle for
   - correct pressure
   - signs of wear
   - evidence of damage
   - safe tread depth
   - objects between twin tyres
   - indications of overheating.
2. Safe driving principles that will help to prevent skids occurring, and the action to take if they do occur.
3. How to drive when the road is
   - icy or snow-covered
   - flooded
   - covered by excess surface water, loose chippings or spillages.
4. What to do if you're involved in a road traffic accident
   - that results in either injury, damage or fire
   - where there's a spillage of hazardous material
   - where danger to other road users results from an obstruction caused by an immobilised vehicle
   - on a motorway.
5. The action to take if your vehicle breaks down during the day time or at night on a
   - bend
   - road with two-way traffic
   - busy dual carriageway
   - clearway
   - motorway
   - railway or LRT crossing

   with particular reference to the safety of passengers.

6. The correct procedure to adopt if an accident occurs that involves a passenger either travelling on your vehicle (e.g., falling over, etc.) or boarding or alighting.
7. The dimensions of your vehicle, including the correct height (especially that of double-decker vehicles, in respect to dangers presented by low bridges, etc.).
8. The weight of your vehicle, in respect to restrictions on weak bridges, etc.
9. The correct procedure to adopt if it becomes necessary to reverse the vehicle while carrying passengers.
10. The differences between toughened and laminated glass used in windows and windscreens.
11. How to use the hammer or similar tool to exit from the vehicle in an emergency.
12. Basic first aid for use on the road.
13. The correct legal procedure (defined in the 1990 amendments to the 1981 PSV regulations) to be adopted by the driver or, where present, a conductor or courier, in respect to any passenger(s) whose behaviour or condition affects the
    - safety of other passengers
    - comfort of other passengers
    - safety of the crew.
14. The appropriate action to take when handed or when finding
    - any lost property
    - suspicious packages.
15. The correct action to take in the event of a passenger, or intending passenger, attempting to alight from or board a moving vehicle.
16. How and when to use fire extinguishers fitted to the vehicle.
17. How to evacuate a PCV when necessary.

TEST PREPARATION

18. How and when to use emergency radio and public address systems, if fitted.

You must appreciate

19. The importance of avoiding any action that could cause offence or provoke physical retaliation.
20. The need to keep control of the permitted number of standing passengers – especially at peak travel times.
21. The need to use safe driving techniques and to obey all speed limits when attempting to maintain schedules laid down in the operator's timetable.
22. The principles of passenger care, including how to
    - communicate effectively
    - assist passengers with special needs
    - help passengers unfamiliar with the service.
23. The importance of presenting a positive image of your company and the industry through your appearance and conduct and the condition of your vehicle.

You must be able to

24. Make a written report, promptly, detailing any defects or symptoms of defects that could adversely affect the safe operation of vehicles. You should submit it to the designated person (the recommended system requires, where practical, a daily 'nil' return to be made to ensure that checks are made).
25. Appreciate when defects are serious enough to require an unroadworthy vehicle to be removed from service.
26. Judge whether a defect is serious enough to cause a vehicle to be unsafe to be driven at all.

Motorway driving

You must have a thorough practical knowledge of the special

- rules
- regulations
- driving techniques

that apply to motorways. In particular, you should know about

- overtaking
- exercising lane discipline
- lanes that are prohibited to certain PCVs
- when speed limiters affect driving
- where PCV speed limits differ to those applying to other traffic
- where temporary speed limits apply when joining and leaving motorways
- breakdowns and emergencies
- driving in all weather conditions
- the principal causes of accidents on motorways.

Safe working practices

You should

1. Know the risks involved in jumping down from cabs (where applicable) and avoid them.
2. Ensure that all doors are closed before the vehicle is moved.
3. Follow safety guidelines when operating
   - under
     - raised engine cowlings
     - raised luggage compartment hatches
     - overhead cables
     - any vehicle

TEST PREPARATION

- near
  - inspection pits
  - wheelchair lift controls
  - refuelling points
  - parked vehicles (especially those likely to be moved or with air suspension)
- whilst
  - carrying out roadside repairs
  - inflating tyres
  - near any vehicle supported on jacks
  - refuelling
  - topping up oil or water.
4. Wear protective clothing, including gloves, when
   - refuelling
   - topping up oil or water
   - checking battery levels
   - emptying waste systems.
5. Know where company policy permits the driver to carry out minor repairs, but do so only
   - if you fully understand how to locate the fault and are able to put it right properly
   - if you can do so without endangering yourself or others
   - with the aid of appropriate equipment, if it's needed
   - if you're sure that any work you do won't invalidate any manufacturer's warranty.

If in doubt, refer to your company.

Revised legislation

Buses carrying children must display a distinctive yellow reflective sign on the front and rear, unless running a scheduled service for use by the general public. This need only be displayed during the morning and afternoon journeys between school and home. Buses displaying the sign are permitted to use hazard warning lights when stationary and when children are boarding or alighting.

Since February 1998

- all minibuses
- all coaches

must have seat belts fitted if they're used for the carriage of three or more children on an organised trip.

The '3 for 2' concession, which allowed three children under 14 years to sit in seats fitted with two seat belts, has been discontinued.

See page 32 for more information on seat belt requirements.

Part Six

The PCV driving test

The topics covered

- What to expect on the day
- Safety checks
- The reversing exercise
- The braking exercise
- The vehicle controls
- The gear-changing exercise
- Other controls
- Moving off
- Using the mirrors
- Giving signals
- Acting on signs and signals
- Making progress
- Controlling your speed
- Separation distance
- Awareness and anticipation
- Hazards
- Selecting a safe place to stop
- Uncoupling and recoupling
- Understanding the rules
- The test results

THE PCV DRIVING TEST

What to expect on the day

Arrive in good time for your test, otherwise it may not go ahead and you'll lose your fee.

The test will last about 90 minutes, so make sure that you won't exceed the number of hours that you're allowed to drive by law, and that you have sufficient fuel.

When you meet the examiner you'll be asked to sign a declaration that the vehicle you're using for the test is fully insured for that purpose.

Your licence

Make sure you bring

- your full car licence (category B)
- the appropriate PCV licence entitlement
  - provisional or
  - full PCV for a category that gives you provisional entitlement
- your theory test pass certificate, if appropriate.

Whatever licence(s) you bring must be signed.

If you don't have your valid, signed driving licence(s) with you the examiner may not be able to conduct your test.

Photographic identification

If your licence does not show your photograph you must also bring with you a form of photographic identification. The examiner will accept

- a signed passport or document of like nature. The passport does not have to be a British one
- any of the following identification cards, provided they have your photograph and your signature
  - workplace identity card
  - trade union or students' union membership card
  - a card for the purchase of reduced price rail tickets
  - cheque guarantee or credit card with photo insert
  - gun licence
- a photograph of yourself, which has been signed and dated on the back by an acceptable person, confirming that the photograph is a true likeness of you. A signature will be accepted from the following
  - LGV trainer on the DSA Voluntary Register of LGV instructors
  - an Approved Driving Instructor (not a trainee)
  - DSA certified motorcycle instructor
  - Member of Parliament
  - local authority councillor
  - teacher (qualified)
  - Justice of the Peace
  - civil servant (established)
  - police officer
  - bank official
  - minister of religion
  - barrister or solicitor
  - medical practitioner
  - commissioned officer in HM forces.

Your test will be cancelled if you can't provide one of these. If you have any queries about what photographic evidence is acceptable, contact the national enquiry line.

No photo

No licence

No test

171

THE PCV DRIVING TEST

Preparing your vehicle

To avoid wasting your own time and the examiner's, make sure that your test vehicle

- has no passengers
- is in the category in which you wish to hold a licence
- doesn't exceed 18.28 metres (60 feet) in length
- has L plates visible to the front and rear (D plates, if you wish, in Wales)
- isn't being used on a trade licence or displaying trade registration plates
- has a secure seat for the examiner, from which he or she can observe the driver
- is fully covered by insurance for its present use and for you to drive
- is legally roadworthy
- has enough fuel, not only for the test (at least 20 miles) but also for you to return to base

Make sure that your vehicle is in a thoroughly roadworthy condition, especially

- stop lamps
- direction indicators
- lenses/reflectors
- mirrors
- brakes
- tyres
- exhaust/silencer
- windscreen/washer/wipers.

It would be unusual for your vehicle not to meet the above requirements, but where vehicles have been adapted for other purposes they may not be suitable for the purposes of the test. If you're in doubt, ask DSA.

You'll be asked to carry out a gear-changing exercise during the test, unless your vehicle is fitted with automatic transmission. Some modern vehicles with automatic and semi-automatic gear-shifting systems may not be suitable for a manual test. The driver taking a manual test must be able to select the gears requested by the examiner and is required to use a clutch pedal whilst moving off, stopping and changing gear.

THE PCV DRIVING TEST

Legal requirements

On the road, the drive will include a gear-changing exercise and the route will cover a wide variety of road and traffic conditions. The route will take in roads carrying two-way traffic, dual carriageways and, where possible, one-way systems.

You'll be expected to demonstrate that you can move off smoothly and safely, both uphill and downhill, in addition to moving off normally ahead and at an angle.

You'll also need to show that you can safely

- meet other vehicles
- overtake
- cross the path of other vehicles
- keep a safe separation distance
- negotiate various types of roundabouts
- exercise correct lane discipline
- display courtesy and consideration to other road users, especially
  - pedestrians
  - riders on horseback
  - cyclists
  - motorcyclists
- apply the correct procedure at
  - pedestrian crossings
  - level crossings (both railway and tramway, where appropriate)
  - traffic signals
  - road junctions.

You'll need to show

- effective use of the mirrors
- correct use of signals
- alertness and anticipation
- correct use of speed
- observance of speed limits
- expert use of the controls.

At the end of your test you'll be asked questions on vehicle safety.

For fuller details of test requirements see page 148.

173

THE PCV DRIVING TEST

Preliminaries

The examiner won't conduct an eyesight test at the start of your test because you will have already met the eyesight and medical requirements before your PCV provisional entitlement was granted.

Before you start the engine

The examiner expects that you've checked and prepared your bus for driving and for taking the test. Before you start your engine you must always be sure that

- all doors are properly closed
- your seat is correctly adjusted and comfortable, so that you can reach all the controls easily and have good all-round vision
- your driving mirrors are correctly adjusted
- your seat belt is fastened, correctly adjusted and comfortable, if fitted
- the handbrake is on
- the gear lever is in neutral.

It's best to develop good habits and to practise this routine while you're learning. The examiner won't be impressed if you have to make adjustments during the test that should have been carried out before it began.

After you start the engine

Don't attempt to drive a vehicle fitted with air brakes until the gauges show the correct pressure or when any warning device (a buzzer sounding or a light flashing) is operating.

If you're driving a vehicle with automatic transmission, you should make sure that the safety checks which apply to your vehicle have been carried out.

174

THE PCV DRIVING TEST

Safety checks

From 1 September 2003, the examiner will ask you to demonstrate or explain how to carry out safety checks on your vehicle before driving. If you are taking a test in a rigid vehicle, the examiner will ask you to demonstrate or explain how to carry out five separate checks. If you are taking a test in a vehicle towing a trailer, the examiner will ask you to demonstrate or explain how to carry out two separate checks.

Skills you should show

You will be expected to know how to check that

- your tyres are correctly inflated, have a safe tread depth and are generally safe to use on the road
- your brakes are working effectively and the pedal does not have excessive travel
- your vehicle has sufficient oil, coolant and hydraulic fluid
- you have sufficient windscreen washer fluid
- the power-assisted steering is working and that excessive 'play' is not apparent
- your headlights, tail lights and reflectors are working and clean
- your brake lights are working and clean
- your horn is working
- the wheel nuts and mudguards are secure
- the vehicle has sufficient air pressure
- the kneeling bus device, if fitted, is working correctly
- the service doors and emergency exits are operating correctly
- all cargo doors are secure if towing a trailer.

You will also be expected to know how to

- check for air leaks
- replace the tachometer disc
- check the windscreen wipers for wear and that the windscreen is clean
- check the suspension for defects
- check the location of first aid equipment, fire extinguishers and other safety equipment
- load a trailer safely
- ensure that the load is secure.

Faults to avoid

You should avoid

- being unfamiliar with the vehicle you are using on test
- being unable to explain or carry out safety checks on the vehicle you are using on test.

THE PCV DRIVING TEST

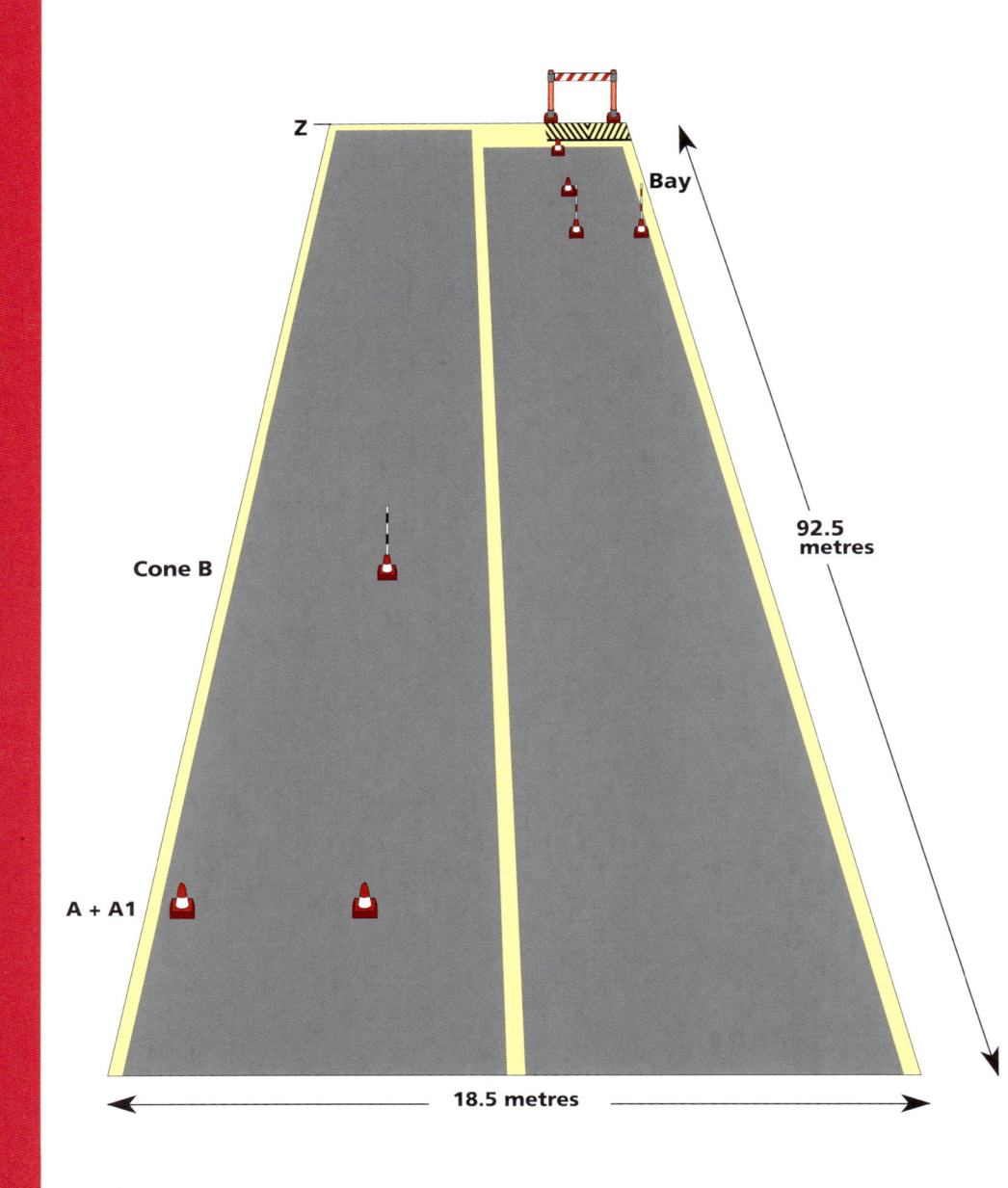

Z

Bay

92.5 metres

Cone B

A + A1

18.5 metres

THE PCV DRIVING TEST

The reversing exercise

You'll be asked to carry out an off-road reversing exercise at the start of the test, whether your test vehicle's coupled to a trailer or not. The examiner will use a diagram of the manoeuvring area to explain the exercise to you. If you take the test with a delegated company examiner, the reversing exercise may consist of reversing into side roads on the left or right during the on-road part of the test.

The diagram opposite shows the area layout for this exercise. From 1 September 2003, the stopping area will have both a solid yellow line and a yellow and black hatched section, and a barrier will be situated at the end of the reversing bay. These changes will not affect the PCV or minibus reversing exercise.

Starting from a fixed point (cones A and A1), you must keep your vehicle inside a clearly defined yellow boundary line so that

- the offside of your vehicle clears cone B
- you stop with the extreme rear of your vehicle in the 1 metre wide yellow/yellow and black stopping area.

At some centres there's also a steel barrier along part of the boundary.

For vehicles without a significant front overhang, cone A is positioned on the area boundary line. For vehicles with a front overhang, the examiner has the discretion to move the cones. If the front axle is well back from the front of the bus or if it has a limited turning circle, cones A, A1 and B may be moved 1 metre (about 3 feet) further into the area from the boundary line.

The distances

The manoeuvring area is 92.5 metres long by 18.5 metres wide (about 300 feet by 60 feet). The overall length for the manoeuvre will be five times the length of the vehicle.

A to A1 = 1½ times the width of the vehicle.

A to B = 2 times the length of the vehicle.

B to line Z = 3 times the length of the vehicle.

The width of the bay will be 1½ times the width of the vehicle. The length of the bay will be based on the length of the vehicle. This can be varied at the discretion of the examiner so that the bay is one of the following lengths

- 1 metre (about 3 feet) longer than your vehicle
- the same length as your vehicle
- 1 metre shorter than your vehicle
- 2 metres (about 6 feet) shorter than your vehicle.

You won't be told the precise length of the bay, as part of this exercise is designed to assess your judgement of the size of your vehicle.

THE PCV DRIVING TEST

What the test requires

The exercise is designed to test your ability to manoeuvre your vehicle in a confined space. You must avoid the marker posts and reverse into a clearly defined bay

- under control
- with reasonable accuracy
- with effective observation throughout.

Skills you should show

The examiner will ask you to drive your bus from where you parked it up to cones A and A1. When he or she signals you to do so, drive up to the cones and stop so that

- the front of your bus is between but not beyond the cones
- the bus is more or less parallel with the yellow boundary line.

If you don't position the vehicle correctly the examiner may ask you to re-position it.

When you're asked you should then

- steer so that the offside of your bus passes clear of cone B (which has a marker pole)
- reverse across the area at a reasonable pace until the rear of your bus enters the bay formed by cones (the two cones at the entrance to the bay will have marker poles)
- carefully control your use of the accelerator, clutch and footbrake throughout
- steer to position your vehicle accurately
- take effective observation throughout the exercise
- make smooth continuous progress across the area
- stop in the position explained to you by the examiner.

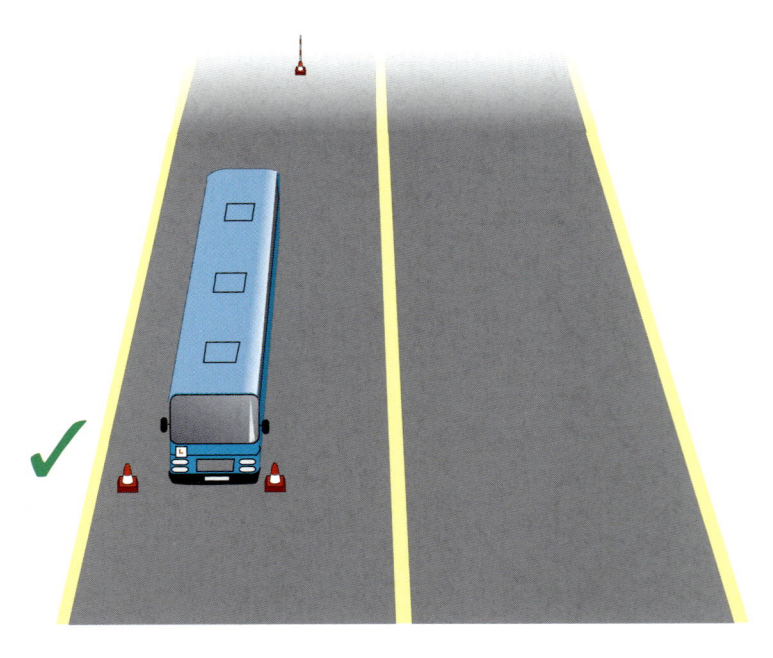

THE PCV DRIVING TEST

Faults to avoid

You should avoid

- approaching the starting point too fast
- not driving in a reasonably straight line as you approach
- stopping beyond the first marker cones A and A1
- turning the steering wheel the wrong way; turning too much or not enough when starting to reverse
- over-steering so that the front offside wheel travels outside the yellow boundary line of the area
- not taking effective observation or misjudging the position of your vehicle so that it hits (or is about to drive over) cone B and its marker pole
- not taking effective observation or misjudging the position of your vehicle so that it hits (or is about to drive over) the cones or marker poles marking out the bay
- allowing the wheel(s) of your vehicle to ride over the boundary lines of the bay or the area
- incorrect judgement so that the rear of your vehicle, and trailer if applicable, is either short of or beyond the yellow/yellow and black stopping area
- taking excessive steering movements or 'shunts' to complete the manoeuvre. Since an overall high standard is expected, only a minimal number of shunts will be accepted
- driving down the area ahead of a position level with cones A and A1 when you're 'shunting' (this is because you'll have gone outside the limits set for your vehicle, i.e., five times its length, after starting the exercise)
- carrying out the manoeuvre at an excessively slow pace
- leaving the cab in order to satisfy yourself of the vehicle's position.

You should remember that throughout the test the examiner will be looking for effective observation and expert handling of the controls.

THE PCV DRIVING TEST

The braking exercise

This exercise usually takes place on the special manoeuvring area at the test centre and not on the public roads. If your test is conducted by a delegated examiner, the braking exercise may be carried out on a quiet public road. The examiner will make sure that no other traffic is close enough to be inconvenienced.

You should make sure that, before you come to the test, there's no loose equipment in the interior of the bus or in luggage lockers. This could fly about and cause injury or damage during the exercise.

The delegated examiner will be with you in the vehicle for this braking exercise. He or she will explain to you the signal to stop. Make sure that you clearly understand what it will be. The signal used will depend on the type of vehicle – it may be the examiner saying 'Stop!' loudly or it may be a bell signal.

In tests conducted by a DSA examiner, the braking exercise is carried out at the LGV/PCV test centre.

What the test requires

Two marker cones approximately 61 metres (200 feet) ahead will be pointed out to you. You should build up the speed of the vehicle to about 32 kph (20 mph). Only when the front of your vehicle passes between the two markers should you apply the brakes.

You should stop

- quickly
- safely
- under full control.

Skills you should show

Stopping the vehicle

- as quickly as possible
- under full control
- as safely as possible
- in a straight line.

THE PCV DRIVING TEST

Faults to avoid

You should avoid

- driving too slowly – less than 32 kph (20 mph)
- braking too soon (anticipating the marker points or the stop signal)
- braking too harshly, causing loss of control
- depressing the clutch too late and stalling the engine
- depressing the clutch well before the brake.

Note
For vehicles fitted with ABS, please refer to the vehicle handbook.

If your bus is fitted with any additional braking controls, such as a retarder, exhaust brake or emergency air brake, you aren't expected to use them in this exercise. This is a test of your ability to stop quickly under normal circumstances.

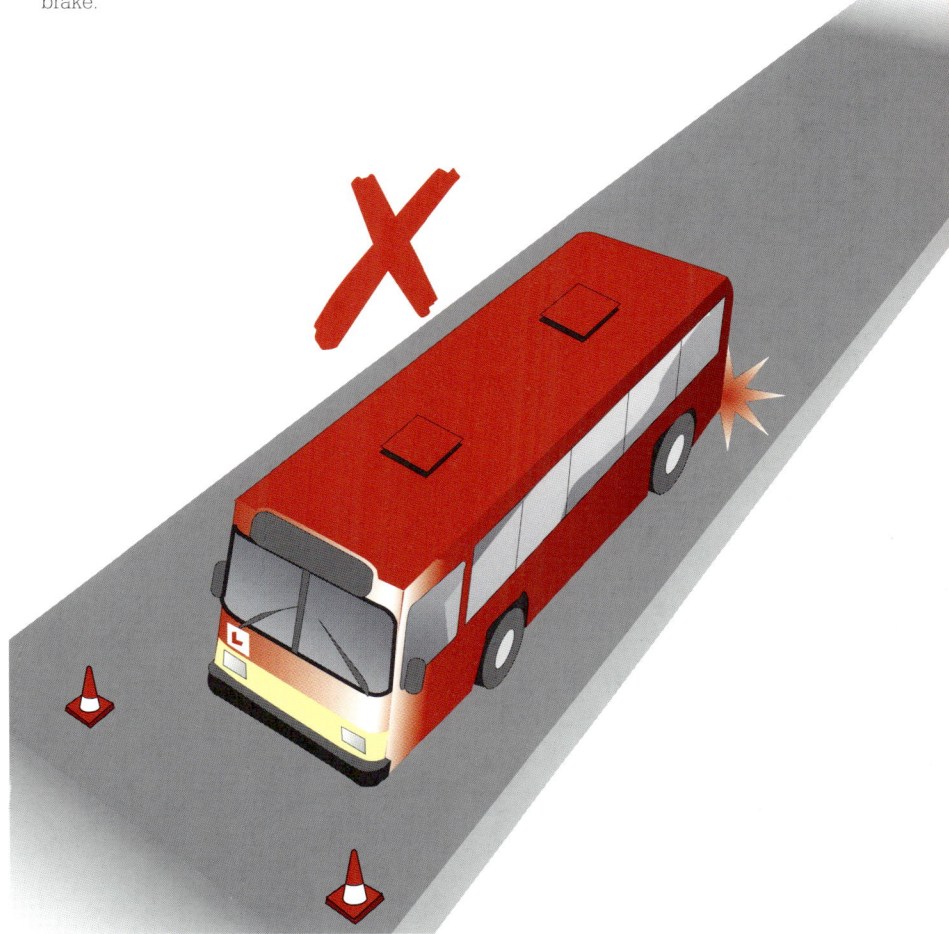

THE PCV DRIVING TEST

The vehicle controls

What the test requires

You must show the examiner that you understand what all the controls do and that you can use them

- smoothly
- correctly
- skilfully
- safely
- at the right time.

In particular, the examiner must be sure that you can properly control the

- accelerator
- clutch
- footbrake
- handbrake
- steering
- gears.

Of course, if your vehicle has automatic transmission some of these won't apply to you. You must

- understand what the controls do
- be able to use them competently.

How your examiner will test you

For this aspect of driving there isn't a special exercise. The examiner will watch you carefully to see how you use these controls. There is, however, a special gear-changing exercise.

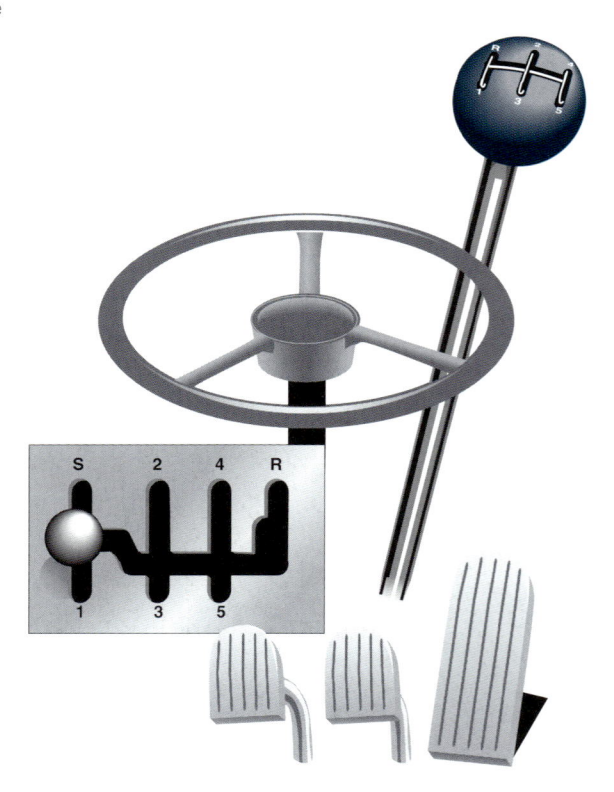

THE PCV DRIVING TEST

Accelerator and clutch

Skills you should show

- Balancing the accelerator and clutch to pull away smoothly.
- Accelerating evenly to gain speed.
- Releasing the accelerator smoothly to avoid erratic driving.
- Depressing the clutch pedal just before the vehicle stops.
- Engaging the clutch smoothly when moving away and changing gear.

Faults to avoid

You should avoid

- loud over-revving, causing excessive engine noise and exhaust fumes. This could alarm or distract other road users
- heavy, inappropriate acceleration followed by immediate braking
- making the vehicle jerk and lurch through uncontrolled use of the accelerator or the clutch
- 'riding' the clutch, that is, failing to take your foot off the pedal when you aren't using it
- jerky and uncontrolled use of the clutch when moving off or changing gear.

183

THE PCV DRIVING TEST

Manual gearboxes

The gears are designed to assist the engine to deliver power under a variety of conditions. The lowest gears may only be necessary if a vehicle is fully loaded or when it's climbing steep gradients.

You should be aware of the manufacturer's advice for the particular vehicle that you drive. Some suggest that first gear should always be used when pulling away, others advise second. Following a manufacturer's advice will minimise clutch and gearbox wear.

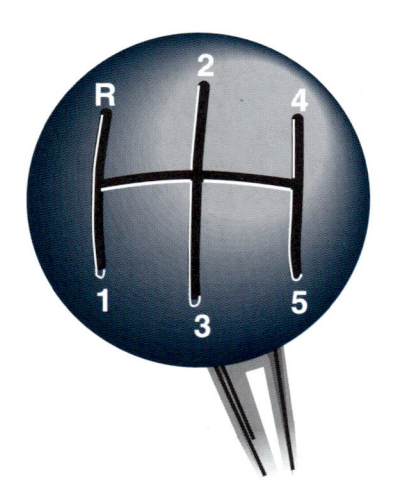

Skills you should show

You should

- move off in the most suitable gear
- choose the most appropriate gear for your speed and the road conditions
- change gear in good time before a hazard or junction
- select the correct gear in good time when climbing or before descending a long hill. On gradients it's essential to plan well ahead.

If you leave gear-changing until you're either losing or gaining too much speed you may have difficulty selecting the gear and maintaining control.

Faults to avoid

You should avoid

- taking your eyes off the road when you change gear
- holding onto the gear lever unnecessarily
- selecting the 'wrong' gear
- 'coasting' with the clutch pedal depressed or the gear lever in neutral.

Coasting is particularly dangerous in vehicles fitted with air brakes. The engine-driven compressor won't replace air being exhausted as the brakes are applied, as it is only running at 'tick over' speed.

184

THE PCV DRIVING TEST

Automatic gearboxes

Modern vehicles may be fitted with sophisticated gearing systems controlled by on-board computers. These systems sense the load, speed, gradient, etc., and select the most appropriate gear for the conditions. On such systems, the driver may only have to ease the accelerator or depress the clutch pedal to allow the system to engage the gear required.

Other automatic gearboxes are controlled by a very simple three-button system

- forward (drive)
- neutral
- reverse.

In spite of this simplicity, it's essential that you learn the correct way to use the system.

Some systems have a 'kick down' facility, whereby a lower gear can be engaged to allow rapid acceleration (e.g., for overtaking). This is achieved by pressing the accelerator to the floor.

Skills you should show

You should

- hold the vehicle firmly on the footbrake before pressing 'forward' to engage the drive. Some systems have interlocks that prevent 'drive' being engaged unless the brake pedal is depressed or the doors are closed, etc.
- press the selector buttons only when the bus is completely stationary
- make careful use of the accelerator to ensure smooth automatic gear-changing.

Faults to avoid

You should avoid

- engaging 'drive' whilst the engine revs are above 'tick over'
- letting the bus remain stationary for long periods with 'forward' or 'reverse' engaged
- forgetting to engage 'drive' before attempting to move off
- not making proper use of any 'kick down' facility.

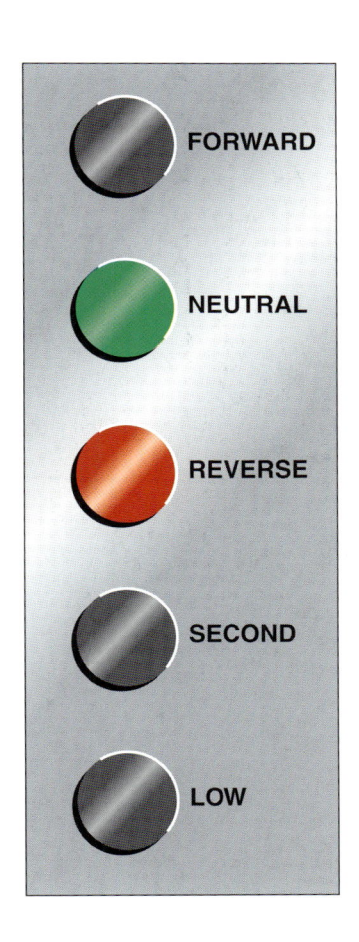

FORWARD

NEUTRAL

REVERSE

SECOND

LOW

185

THE PCV DRIVING TEST

Semi-automatic gearboxes

The system most commonly found on buses and coaches is one whereby the driver has full control over the gear selected, but has no clutch pedal. This is often a 'pneumo-cyclic' gearbox, which consists of a number of electronic relays controlling air systems that make the actual gear changes. The gears are chosen by means of a 'gate' selector. When coupled with a fluid flywheel, this system eliminates the need for the use of a clutch when pulling away, stopping or changing gear. However, smooth changes require some skill and practice to achieve.

Again, the most appropriate method of changing gear will depend on the manufacturer's advice for the particular vehicle. Most advise that, when changing between one gear and another, a brief pause be made when the lever is in neutral. The accelerator should be depressed to a level appropriate for the gear about to be engaged.

Usually, semi-automatic gearboxes are coupled to diesel engines. Thus the amount of time needed in neutral to allow the engine revs to match the road speed needs careful consideration.

Skills you should show

You should

- hold the vehicle firmly on the footbrake before engaging forward or reverse gears from a standstill. Some systems have interlocks that prevent 'drive' being engaged unless the brake pedal is depressed or the doors are closed, etc.
- make careful use of the accelerator to ensure smooth gear-changing.

Faults to avoid

You should avoid

- engaging a forward or reverse gear from a standstill whilst the engine revs are above 'tick over'
- letting the bus remain stationary for long periods with a forward or reverse gear engaged
- forgetting to engage a gear before attempting to move off
- not making proper use of the gear selector and accelerator.

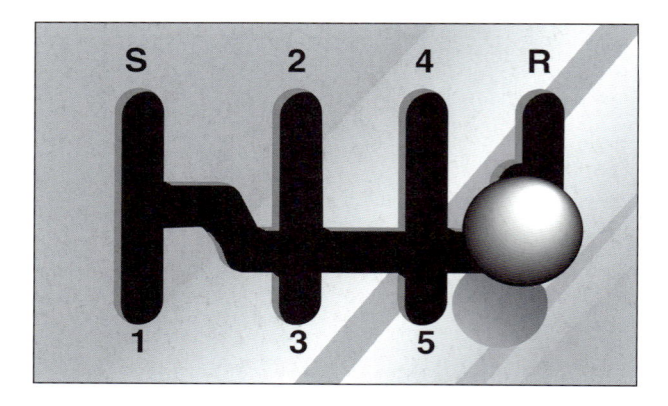

THE PCV DRIVING TEST

The footbrake

With all braking systems it is important to remember that there is a direct relationship between the pressure applied to the footbrake pedal and the braking force exerted on the wheels.

Older vehicles may have vacuum brake systems, which have similar characteristics to air brakes.

Some other vehicles may have a system known as 'air over hydraulic', in which air pressure operates a hydraulic braking system. These are usually lighter vehicles and the system is designed to make the brakes less harsh. Remember that controlled progressive braking is required at all times.

Skills you should show

You should

- brake in good time
- brake lightly in most situations
- brake progressively
- use the correct technique for releasing pressure on the brake just before coming to rest. This allows you to stop the bus without jerks.

Faults to avoid

You should avoid

- braking harshly
- excessive and prolonged use of the footbrake
- braking and steering at the same time, unless you're already travelling at low speed
- braking in a way that would cause passengers discomfort.

THE PCV DRIVING TEST

The handbrake

Buses, coaches and minibuses are equipped with one of two types of handbrake.

Mechanical

- Generally found on older or smaller vehicles.
- Comprising a long lever with a button, or more usually, a squeeze-grip release, as in a car. The lever pulls a series of cables that apply the rear (or, more rarely, the front) brakes.

Air-operated

- Fitted to vehicles with air or 'air over hydraulic' footbrake systems.
- Operated by a small lever with a collar or a push-button release.

Skills you should show

You should

- know how and when to apply the handbrake
- apply the handbrake before leaving the cab when you intend to secure the vehicle
- co-ordinate your use of the handbrake and other controls in order to achieve smooth uphill starts.

Some modern braking systems will apply a parking brake when the vehicle is brought to a stop by the footbrake. The handbrake is released in the normal way. You should know how to operate this system if it's fitted to a vehicle that you intend to drive.

Faults to avoid

- applying the handbrake before the vehicle has stopped
- attempting to move off with the handbrake still applied
- using the 'park' position on the gear selector on automatic vehicles as a substitute for applying the handbrake. Only the gearbox locks when 'P' is engaged – the vehicle may be free to move when the next driver selects neutral to start the engine
- 'holding' the vehicle on the clutch on uphill slopes (in manual buses and coaches). This can cause excessive clutch wear.

The clutch isn't designed to prevent a vehicle weighing up to 18 tonnes from moving backwards. You should always apply the handbrake and carry out the correct uphill start procedure to avoid unnecessary wear and tear on the clutch.

THE PCV DRIVING TEST

Emergency brakes

All vehicles are required to have at least two braking systems so that failure of one won't prevent the vehicle being brought to rest safely.

'Split' systems are often fitted to ensure that failure of one part of the normal braking system leaves other parts operational. 'Fail safe' systems can result in the automatic gradual application of all or some of the brakes if the driver ignores brake warning indicators.

Construction and Use regulations do require the driver to be able to apply all or part of the braking system in the event of footbrake failure. In simple terms this means that, if you press the footbrake and nothing happens, you must have another means of stopping. How you do this depends on the system fitted

- mechanical handbrake: the handbrake can be applied progressively to bring the vehicle to a stop
- emergency brake: a separate lever is provided on some older air- and vacuum-braked buses to allow progressive application of the brakes
- air-operated handbrake: the brake is partially applied to allow progressive application of the brakes.

During your training you should find out which method should be used with the vehicle that you drive. You should practise it in a safe place, preferably off the road and under expert supervision.

You must not use this method of braking at any time during your test.

Skills you should show

The braking exercise in the driving test requires a rapid, controlled stop, using the footbrake. It isn't an emergency stop exercise.

Normally it won't be necessary to demonstrate any emergency braking systems.

Faults to avoid

You should avoid

- immediate full application of the brake
- locking the wheels and skidding
- coming to rest heavily in a way that may injure passengers.

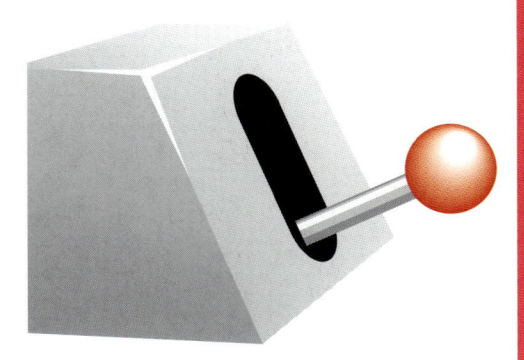

189

THE PCV DRIVING TEST

Steering

The bus, coach or minibus you drive will probably have power-assisted steering. With power assistance the steering effort required is greatly reduced through the action of an engine-driven pump.

It's generally necessary to take corners more slowly when you don't have the benefit of power assistance, simply because the gearing at the steering wheel is lower and it takes more effort and time to turn it. The danger with power assistance is that the lack of effort required (and, in some cases, the lack of 'feel' transmitted back to the driver) can result in taking corners too quickly. This can put either safety or comfort at risk. You need to be aware of this.

Skills you should show

You should

- place your hands on the steering wheel in a position that's comfortable and which gives you full control at all times
- keep your steering movements steady and smooth
- steer an accurate path and be aware of the 'swept path' that your vehicle will take.

It's particularly important to take the correct path when driving a bus with long overhangs or limited ground clearance.

Faults to avoid

You should avoid

- turning the wheel too early when turning a corner. You risk
  - cutting the corner when turning right, causing the rear wheel(s) to cut across the path of traffic waiting to emerge
  - striking the kerb when turning left
- turning too late. You could put other road users at risk by
  - swinging wide at left turns
  - overshooting right turns
- crossing your hands on the steering wheel (whenever possible)
- allowing the wheel to spin back after turning
- resting your arm on the door.

Remember, the stability of a bus can be affected by cornering too quickly.

190

THE PCV DRIVING TEST

The gear-changing exercise

How your examiner will test you

The examiner will ask you to pull up at a convenient place to carry out the gear-changing exercise (usually at an early stage in the test).

You'll be asked to move off in the lowest gear and change up to each gear in turn until you reach a gear the examiner considers appropriate for your vehicle. This will depend on the gearbox fitted. The examiner will make sure that you understand which gear should be reached.

When you've reached the agreed gear the examiner will want to see you change back down into each gear in turn, driving for a short distance in each gear until you reach the lowest gear again.

You won't be asked to use anything other than the normal gears appropriate to your vehicle.

The gear-changing exercise enables the examiner to see whether you engage the lower gears competently. An opportunity for you to demonstrate this skill may not occur naturally during the test.

Skills you should show

You should

- move off smoothly in the lowest gear
- change up to the next gear as soon as the correct speed is reached
- show smooth, unhurried and precise gear changes
- match the road speed of the vehicle to the lower gear, when changing down, by careful use of the footbrake, if necessary.

Throughout this exercise you must

- use effective observation, especially before moving off and slowing down
- give any signal that may be appropriate
- use the controls skilfully to ensure smooth engagement of the gears.

Faults to avoid

You should avoid

- not showing sufficient consideration for other road users before moving off, during the exercise or when slowing down by
  - forgetting to check the mirrors
  - not acting sensibly in response to the actions of other drivers
  - not signalling to inform other road users before moving off or slowing down
- jerky use of the accelerator or clutch
- not starting off in the lowest gear
- not selecting the next gear in sequence
- not being able to engage a gear
- not slowing the vehicle down enough before selecting a lower gear.

THE PCV DRIVING TEST

Other controls

You should understand

- The functions of all controls and switches that have a bearing on road safety, for example
  - indicators
  - lights
  - windscreen wipers
  - demisters.
- The meaning of gauges or other displays on the instrument panel, especially
  - air pressure gauge(s)
  - speedometer
  - various warning lights/buzzers
  - on-board computer displays
  - braking systems failure warnings
  - bulb failure warnings
  - gear-selection indicators.
- Time, speed and distance recording equipment, including
  - operating tachographs
  - completing tachograph charts
  - keeping records
  - the operation of any speed-limiting equipment fitted.

Safety checks

You should also be able to

- carry out routine safety checks on
  - oil and coolant levels
  - tyre pressures
- identify defects, especially with
  - steering
  - brakes
  - tyres
  - seat belts
  - lights
  - reflectors

- horn
- rear view mirrors
- speedometer
- exhaust system
- direction indicators
- windscreen, wipers and washers
- wheel-nut security
- understand the effects that any fault or defect will have on the handling of your vehicle.

Warning

Some bus manufacturers, but not all, fit wheel nuts that tighten clockwise on the nearside of the vehicle and anti-clockwise on the offside. Make sure that you know which thread is fitted to your vehicle before you attempt to tighten them. The consequences of getting it wrong are dangerous. In any case, it's much better to entrust this to trained mechanics, wherever possible, as the nuts should always be tightened to the specified torque. Some wheels are 'spigot-mounted' and require specialised knowledge when being removed or refitted.

THE PCV DRIVING TEST

Moving off

What the test requires

From a standstill, you must be able to move off safely and under control

- on the level
- from behind a parked vehicle
- on a hill
- uphill and downhill.

How your examiner will test you

The examiner will watch your observation and use of the controls each time you move off.

Level and uphill starts

- Aim to co-ordinate your use of the controls so that the vehicle remains stationary momentarily when the handbrake is released, ready to move off.
- Check all round for pedestrians and other road users. Move off if it's clear.
- If there's more than a moment's delay between releasing the brake and moving off, reapply the handbrake and repeat the sequence when it's safe to do so.

Downhill starts

- Prevent the vehicle from moving when you release the handbrake by applying the footbrake first.

Angle starts

- Ensure that you apply sufficient steering to pass the parked vehicle safely.
- Ensure that you won't endanger traffic when you move away.

THE PCV DRIVING TEST

Using the mirrors

Mirrors are one of the best aids to road safety. They help you to avoid causing problems to other road users and allow you to predict when to take action safely.

Try not to think of mirror checks as something you do because you've been told to. The important point isn't that you've looked in the mirrors, but that you've gained additional information to help you to drive safely.

Try to time your mirror checks to allow time to assess what you see before taking any action.

Sequence of mirror checks

Professional drivers develop a technique for checking mirrors whilst remaining fully aware of what's happening ahead. Whatever method you adopt, the examiner will observe how you use the mirrors and whether you act sensibly on what you see.

When you're on the road hazards often occur together, or one immediately after another. One may also happen just when the need to begin a manoeuvre to deal with another occurs.

You must ensure that you observe every potential danger and are fully prepared to deal with it, if it occurs.

The sequence of checks has to be adapted as situations develop. In reality, it takes only moments to carry out and should become second nature without the need to constantly analyse what you're doing.

Order of checks

- Identify the hazard that gives rise to the need to manoeuvre.
- Assess where the greatest potential danger lies in the intended manoeuvre – to the right or to the left of your vehicle. Check that mirror first.
- Check the mirror on the other side.
- As your eyes return to the road ahead reassess the hazard and, if you have an interior mirror that allows you to see what's happening behind, check the position of following traffic.
- Check the first mirror again and, as your eyes return to the road ahead, assess what you've seen and signal if necessary.
- Carry out the manoeuvre, if it's still safe to do so, rechecking the mirrors as necessary.

THE PCV DRIVING TEST

Example: moving out to pass a parked car

You see a parked car some distance ahead.

- The primary danger is that someone may attempt to overtake you as you need to move out. Check the offside mirror.
- Check the nearside mirror.
- Check that the parked car hasn't moved away, or that you'll need to give it extra room because someone's about to get out.
- Will you need to wait for approaching traffic?
- Look to see what traffic is following you by checking the offside mirror.
- Has the situation ahead changed? Signal right, if necessary.
- Begin to move out if it's safe to do so, or wait if it isn't.
- Keep checking how the situation develops.

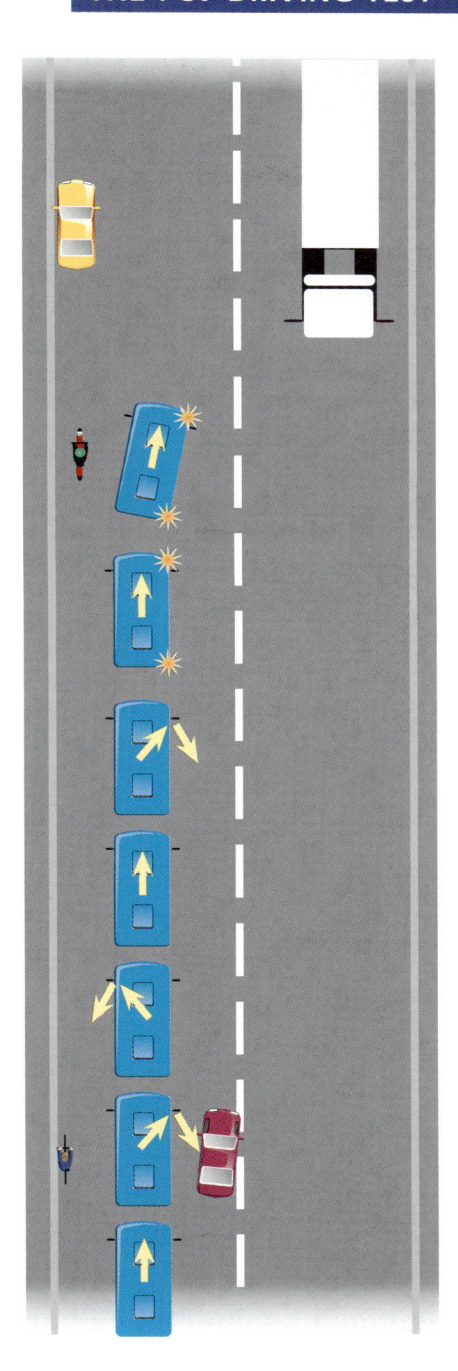

THE PCV DRIVING TEST

What the test requires

Make sure that you use your mirrors effectively

- before any manoeuvre
- to keep up to date on what's happening behind you.

Check carefully before

- moving off
- signalling
- changing direction
- turning left or right
- overtaking or changing lanes
- increasing speed
- slowing down or stopping
- opening any offside door.

Check again in the nearside mirror after

- passing parked vehicles
- passing horse riders, motorcyclists or cyclists
- passing any pedestrians standing close to the kerb
- passing any vehicle you've just overtaken and before moving back to the left.

How your examiner will test you

For this aspect of driving there isn't a special exercise. The examiner will watch your use of mirrors as you drive.

Skills you should show

You should

- establish good habits by
  - looking before you signal
  - looking and signalling before you act
  - acting sensibly and safely on what you see in the mirrors
- be aware that the mirrors won't show everything behind you
- check your nearside mirror, every time, after passing
  - parked vehicles
  - vulnerable road users
  - vehicles you've just overtaken
- always be as aware of what's happening behind and alongside as you are of what's going on ahead
- always be aware of the effect your vehicle has on any vulnerable road users that you may pass.

THE PCV DRIVING TEST

Faults to avoid

You should avoid

- manoeuvring without checking the mirrors first
- not acting on what you see when you look in the mirrors
- taking action at the same time as looking in the mirrors, instead of as a result of what you see in them
- looking in the mirrors at an inappropriate moment so that you fail to observe changes in the situation ahead.

Remember, just looking isn't enough. Acting sensibly on what you see is more important. Always use the MSM/ PSL routine

- Mirrors
- Signal
- Manoeuvre
  - Position
  - Speed
  - Look.

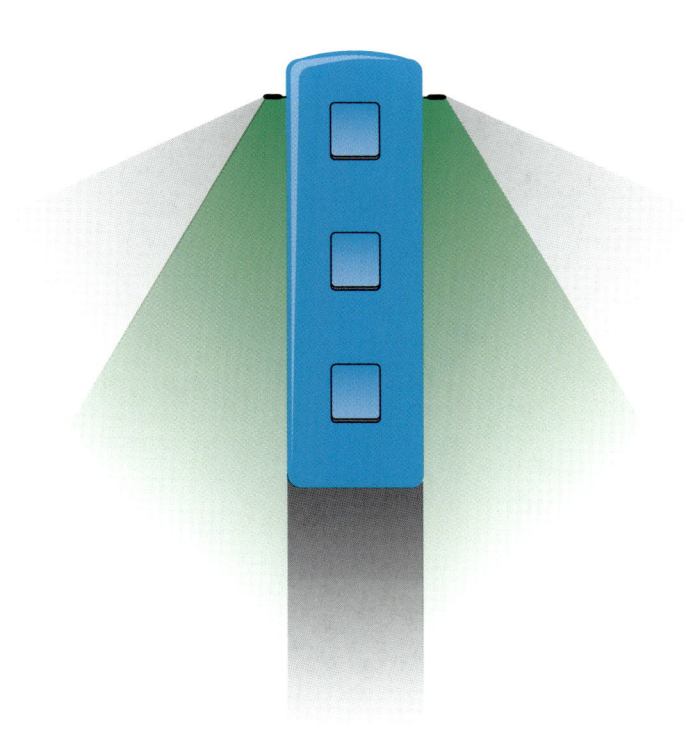

197

THE PCV DRIVING TEST

Giving signals

What the test requires

You must give clear signals in good time so that other road users know what you intend to do next. This is particularly important with long PCVs because other road users may not understand the position you need to move into

- before turning left
- before turning right
- at roundabouts
- to move off at an angle
- before reversing into an opening.

You must only use the signals shown in *The Highway Code*, as any others may be misunderstood.

Any signal you give must help other road users to

- understand what you intend to do next
- take appropriate action.

Always check that you've cancelled an indicator signal as soon as it's safe to do so.

How your examiner will test you

For this aspect of driving there isn't a special exercise. The examiner will watch carefully to see how you use signals in your driving.

Skills you should show

Give any signals

- clearly
- at the appropriate time
- by indicator
- by arm, if necessary.

Make sure that any signal you give is visible long enough for other road users to see it and understand its meaning.

Faults to avoid

You should avoid

- giving misleading or incorrect signals
- omitting to cancel signals
- waving on pedestrians to cross in front of your vehicle (neither you nor they may have seen a dangerous situation, which you're 'inviting' them towards)
- giving signals other than those shown in *The Highway Code*.

198

THE PCV DRIVING TEST

Acting on signs and signals

What the test requires

You must have a thorough knowledge of traffic signs, signals and road markings. You should be able to

- recognise them in good time
- take appropriate action on them.

At the start of the road section of the PCV driving test the examiner will ask you to continue to follow the road ahead, unless traffic signs indicate otherwise or unless you're asked to turn left or right. You'll be given any direction to turn in good time. If you aren't sure, ask the examiner to repeat the direction.

Skills you should show

Traffic lights

You must

- comply with traffic lights
- approach at such a speed that you can stop, if necessary, under full control
- only move forward at a green traffic light if
  - it's clear for you to do so
  - by doing so your vehicle won't block the junction.

Authorised persons

You must comply with signals given by

- police officers
- traffic wardens
- school crossing patrols
- any authorised person controlling traffic, e.g., at road repairs.

Other road users

You must watch for signals given by other road users and

- react safely
- take appropriate action
- anticipate their actions
- give signals to any traffic following your vehicle that may not be able to see the signals given by a road user ahead of you. This is particularly important when a vehicle or rider ahead is intending to turn right, and the size of your vehicle prevents traffic behind you from seeing their signal.

199

THE PCV DRIVING TEST

Making progress

The examiner will be looking for a high standard of driving from an experienced driver displaying safe, confident driving techniques. You aren't a learner driver and you won't pass the test if you drive hesitantly or in a way that shows you aren't fully in control of your vehicle.

Because you're an experienced driver and are expected to drive accordingly, you must

- select a safe speed to suit road, weather and traffic conditions
- move away at junctions as soon as it's safe to do so
- avoid stopping unnecessarily
- make progress when conditions permit.

How your examiner will test you

For this aspect of driving there isn't a special exercise. The examiner will watch your driving and will expect to see you

- making reasonable progress where conditions allow
- keeping up with the traffic flow when it's safe and legal to do so
- making positive, safe decisions as you make progress.

Skills you should show

You should

- drive at the appropriate speed, depending on the
  - type of road
  - traffic conditions
  - weather conditions and visibility
- approach all hazards at a safe speed without being unduly cautious or holding up following traffic unnecessarily.

Faults to avoid

You should avoid

- driving so slowly that you hinder other traffic
- being over-cautious or hesitant
- stopping when you can see that it's obviously clear and safe to make progress.

200

THE PCV DRIVING TEST

Controlling your speed

What the test requires

You should make good progress when possible, taking into consideration

- the type of road
- the volume of traffic
- the weather conditions and the state of the road surface
- the braking characteristics of your vehicle
- speed limits that apply to your vehicle
- any hazards associated with the time of day (school times, etc.).

How your examiner will test you

For this aspect of driving there isn't a special exercise. The examiner will watch carefully your control of speed as you drive.

Skills you should show

You should

- take great care in the use of speed
- drive at the appropriate speed to the traffic conditions
- be sure that you can stop safely in the distance that you can see to be clear
- leave a safe separation distance between your vehicle and the traffic ahead
- allow extra stopping distance on wet or slippery road surfaces
- observe the speed limits that apply to your vehicle
- drive sensibly and anticipate any hazards that could arise
- allow for other road users making mistakes.

Faults to avoid

You should avoid

- driving too fast for the conditions
  - road
  - traffic
  - weather
- exceeding speed limits
- varying your speed erratically
- having to brake hard to avoid a situation ahead
- approaching bends, traffic signals and any other hazards at too high a speed.

THE PCV DRIVING TEST

Separation distance

Always keep a safe separation distance between you and the vehicle in front.

What the test requires

You must always drive at such a speed that you can stop safely in the distance that you can see to be clear.

In good weather conditions, leave a gap of at least 1 metre (about 3 feet) for each mph of your speed, or a two-second time gap.

In bad conditions, leave at least double that distance, or a four-second time gap.

In slow-moving congested traffic it may not be practical to leave as much space, but you must always be sure that you can stop safely – whatever happens.

How your examiner will test you

For this aspect of driving there isn't a special exercise. The examiner will watch carefully and take account of your

- use of the MSM/PSL routine
- anticipation
- reaction to changing road and traffic conditions
- handling of the controls.

Skills you should show

You should

- be able to judge a safe separation distance between you and the vehicle ahead
- show correct use of the MSM routine, especially before reducing speed
- avoid the need to brake sharply if the vehicle in front slows down or stops
- take extra care when your view ahead is limited by large vehicles, such as other buses or lorries.

Watch out for

- brake lights ahead
- direction indicators
- vehicles ahead braking without warning.

Faults to avoid

You should avoid

- following too closely or 'tailgating'
- braking suddenly
- swerving to avoid the vehicle in front, which may be slowing down or stopping
- not leaving side road junctions clear when a queue of traffic stops.

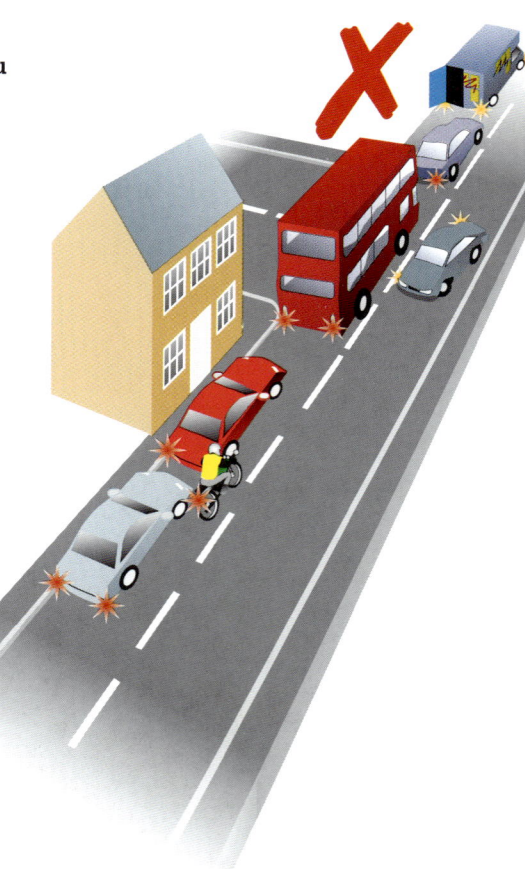

THE PCV DRIVING TEST

Awareness and anticipation

The traffic situation can change from second to second, depending on the time of day, the location and the density of traffic. Sometimes you can see that a situation is obviously going to turn dangerous. The skilful driver anticipates what might happen.

As the driver of a bus, coach or minibus you must constantly drive with this sense of awareness and anticipation. Ask yourself

- what's happening ahead?
- what are other road users doing, or about to do?
- do I need to
  - speed up?
  - slow down?
  - prepare to stop?
  - change direction?

It's essential to be fully alert at all times and to scan the road ahead constantly. By doing this you'll remain in control of both the situation and your vehicle.

In fast-moving traffic you'll need to be constantly checking and re-checking the scene around you. It's essential to recognise well in advance the mistakes other road users may be about to make.

Hazards

What is a hazard?

When you're moving, a hazard is any situation that could involve adjusting speed or altering course. Look well ahead for

- road junctions or roundabouts
- parked vehicles
- cyclists or horse riders
- pedestrian crossings.

By identifying the hazard early enough you'll have time to take the appropriate action.

When you're stationary, a hazard can be created by the actions of other road users around you. Watch for

- pedestrians crossing in front
- cyclists or motorcyclists moving up alongside
- drivers edging up on the nearside before you make a left turn
- vehicles pulling up close behind when you intend to reverse.

Stay on the alert and watch what's happening around you.

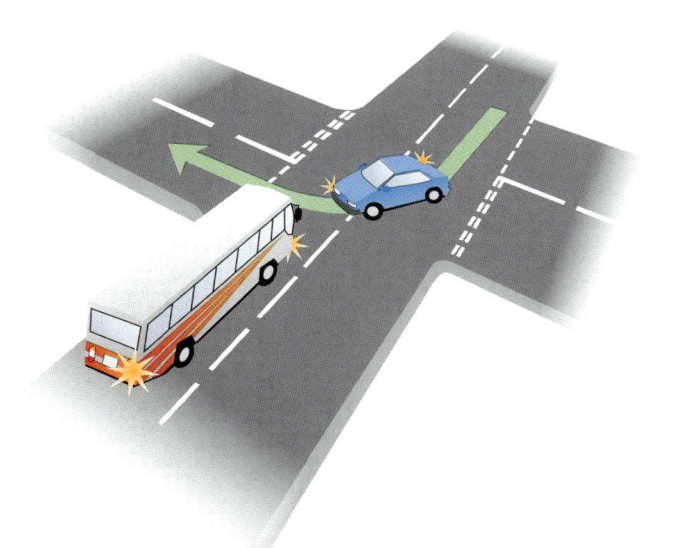

THE PCV DRIVING TEST

Hazards – other road users

Skills you should show

Pedestrians
- Give way to pedestrians when turning from one road into another, or when entering premises such as bus or railway stations, schools, etc.
- Take extra care with the
  - very young
  - disabled
  - elderly

  as they may not realise you won't be able to stop suddenly.

You must be even more vigilant when driving through shopping areas, for example, where there are often large numbers of people waiting to cross at corners. Drive slowly and considerately when you need to enter any pedestrianised areas.

Cyclists
Take extra care when

- crossing cycle lanes
- you can see a cyclist near the rear of your vehicle or moving up along the nearside as you're about to turn left
- approaching any children on cycles
- there are gusty wind conditions.

Motorcyclists
Watch for motorcyclists

- filtering in slow traffic streams
- moving up along the side of your vehicle
- especially when you're about to move out at junctions.

Think once

    Think twice

        Think bike.

THE PCV DRIVING TEST

Horse riders and animals

Remember, the size, noise and sometimes even the colour of your vehicle can unsettle even the best mannered horse. Watch young, possibly inexperienced, riders closely for signs of any difficulty with their mounts. Give horse riders as much room as you can.

Avoid the need to rev the engine until you're clear of the animal. Several light applications of the brakes as you approach should ensure that the air brake system relief valve doesn't blow off just as you're level with the animal.

React in good time to anyone herding animals. Look out for warning signs or signals in rural districts.

Faults to avoid

You should avoid

- sounding the horn unnecessarily
- revving the engine, deliberately
- causing the air brakes to 'hiss' by heavy applications
- edging forward when pedestrians are crossing in front of your vehicle
- any signs of irritation or aggression towards other road users, especially the more vulnerable.

THE PCV DRIVING TEST

Hazards – positioning and lane discipline

What the test requires

You should

- normally keep well to the left
- keep clear of parked vehicles
- avoid weaving in and out between parked vehicles
- position your vehicle correctly for the direction you intend to take
- obey road markings, especially
  - left- and right-turn arrows at junctions
  - when approaching roundabouts
  - in one-way streets
  - bus lanes
  - road markings for PCVs or LGVs approaching arched or narrow low bridges.

With long PCVs, only straddle lane markings or move over to the left or right when necessary to avoid mounting the kerb or colliding with lamp-posts, traffic signs, etc.

How your examiner will test you

For this aspect of driving there isn't a special exercise. The examiner will watch carefully to see that you

- use the MSM routine
- select the correct lane in good time.

THE PCV DRIVING TEST

Skills you should show

You should

- use the MSM/PSL routine correctly
- plan ahead and choose the correct lane in good time
- position your vehicle sensibly, even if there aren't any lane markings.

Always remember that other road users may not understand what you intend to do next. Watch them carefully and ensure that you signal in good time.

Faults to avoid

You should avoid

- driving too close to the kerb
- driving too close to the centre of the road
- changing lanes at the last moment or without good reason
- hindering other road users by being incorrectly positioned or in the wrong lane
- straddling lanes or lane markings when it's unnecessary
- using the size of your vehicle to block other road users from making progress
- cutting across the path of other road users in another lane at roundabouts.

THE PCV DRIVING TEST

Hazards – junctions

The size of your vehicle and the difficulties that may arise when manoeuvring it mean that it's essential to make the correct decisions at road junctions.

Never drive into a situation that you can't see a clear path through. If you drive your vehicle into a blocked road any traffic building up behind will prevent you from reversing out, leaving you in an impossible position. Similarly, if you need to wait for an obstruction to clear, stop in a position that allows you an escape route if at all possible.

What the test requires

You should

- use the MSM/PSL routine in good time on the approach to junctions
- assess the situation correctly, so that you can position the vehicle to negotiate the junction safely
- take as much room as you need on approach to see the road space available. There may not be enough room for a wide swing in the road that you're entering
- take advantage of any improved vision from the driving position in your vehicle and stop or proceed as necessary
- be aware of any lane markings and the fact that your vehicle may have to occupy part of the lane alongside
- try to position in good time in one-way streets
- make sure you take effective observation before emerging at any road junction
- use your mirrors to observe the rear wheels of your vehicle as you drive into and out of the junction
- correctly assess the speed of oncoming vehicles before crossing or entering roads with fast-moving traffic
- always allow for the fact that you'll need time to build up speed in the new road.

If you're crossing a dual carriageway or turning right onto one, don't move forward unless you can clear the central reservation safely. If your vehicle is too long for the gap, wait until it's clear from both sides and there's a safe opportunity to go.

208

THE PCV DRIVING TEST

How your examiner will test you

For this aspect of driving there isn't a special exercise. The examiner will watch carefully and take account of your

- use cf the MSM/PSL routine
- position and speed on approach
- observation and judgement.

As an aid to remembering the correct routine, think of the word LADEN

- Look well ahead on approach
- Assess conditions at the junction
- Decide when it's safe to go
- Emerge from (or enter) the junction safely
- Negotiate the hazard (junction) safely.

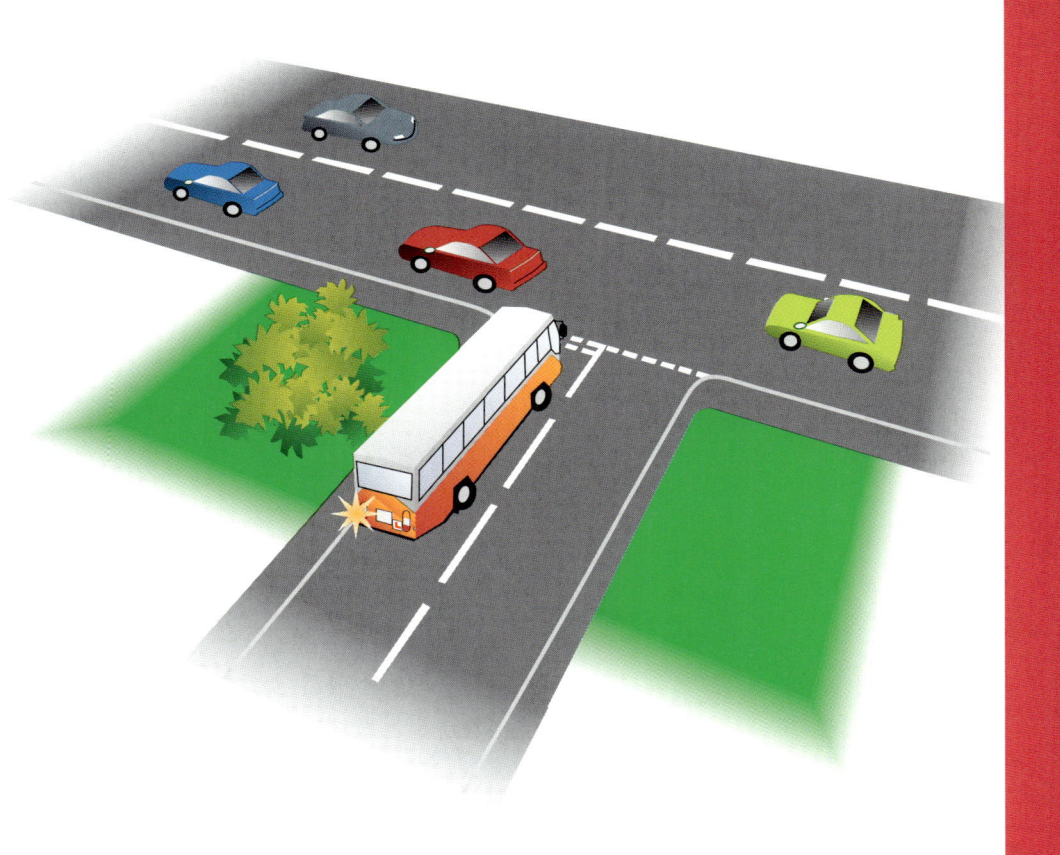

THE PCV DRIVING TEST

Hazards – roundabouts

What the test requires

Roundabouts can vary in size and complexity, but the object of all of them is to allow traffic to keep moving, wherever possible.

Some roundabouts are so complex that they require traffic lights to control the volume of traffic, whilst at others signals operate at peak periods only.

At the majority of roundabouts traffic is required to give way to the vehicles approaching from the right. However, at some locations the 'Give way' signs and markings apply to traffic already on the roundabout. You must be aware of these differences.

Skills you should show

You should plan your approach well in advance and use the MSM/PSL routine in good time. With buses, it's essential to adopt the appropriate lane, depending on the exit you intend to take and the size of your vehicle.

Lane discipline

You should

- plan well ahead
- look out for traffic signs as you approach
- have a clear picture of the exit you need to take
- look out for the number of exits before yours
- either follow the lane markings, as far as possible, or select the lane most suitable to the size of your vehicle
- signal your intentions clearly and in good time
- avoid driving into the roundabout too close to the right-hand kerb
- as it isn't always possible to keep your vehicle within road markings, make frequent mirror checks to ensure that you aren't endangering others
- accurately assess the speed and intentions of traffic approaching from the right.

Always watch any vehicle in front when you're about to enter the roundabout. When you see a gap in the traffic, always check to make sure the vehicle in front of you has pulled off before you do so. Drivers sometimes change their minds at the last moment. Many rear-end collisions take place in just these circumstances.

THE PCV DRIVING TEST

Unless lane markings or road signs indicate otherwise you should follow the procedure noted here when turning left or right, or when going straight ahead.

Turning left
You should

- check your mirrors
- give a left-turn signal in good time as you approach
- approach in the left-hand lane. With a long vehicle you might need to take up some of the lane on your right, depending on how sharp or narrow the exit turn is
- adopt a path that ensures your rear wheels don't mount the kerb
- give way to traffic approaching from the right, if necessary
- use the nearside mirror(s) to be sure that no cyclists or motorcyclists are trapped on the nearside
- use the offside mirror to check that no passing vehicle will be hit when the rear overhang swings out as you begin to turn
- continue to signal through the turn
- look well ahead for traffic islands/bollards in the middle of your exit road, which will restrict the width available to you

Going ahead
You should

- approach in the left-hand lane unless blocked or clearly marked for 'left turn' only
- not give a signal on approach (other than brake lights, if you need to reduce speed)
- try to stay in the lane if possible, depending on the length of your vehicle
- keep checking the mirrors. Be aware that other road users may not anticipate the 'swept path' of your vehicle. Be prepared to stop if they don't, as swerving will normally make matters worse
- indicate left as you pass the exit just before the one that you intend to take
- look well ahead for traffic islands/bollards in the centre of your exit road
- make sure that the rear wheels don't mount the kerb as you leave the roundabout.

THE PCV DRIVING TEST

Turning right or full circle

You should

- look well ahead and use the MSM/PSL routine in good time
- signal right in good time before moving over to the right on approach. Watch for any vehicles, especially motorcycles, accelerating up on the offside of your vehicle
- make frequent mirror checks
- only enter the roundabout when you're sure that it's safe to emerge
- keep checking for traffic coming from your right.

When you need extra space and only one lane is marked for 'right turn', occupy part of the lane to your left. Do this on the approach and through the roundabout.

When driving a long vehicle, if there are two lanes available, use the left-hand of the two lanes.

Don't pull out across the path of any vehicle closely approaching from the right. Not only could the approaching vehicle be travelling at speed, but it could also be moving on a curved course so any sudden braking would be likely to send it into a skid.

You should

- use the mirrors to observe traffic coming round with you on the nearside, and also to check that your rear wheels are keeping clear of the kerb on the roundabout itself
- change your signal to 'left turn' as you pass the exit before the one you wish to take.

This procedure is useful when you need to turn a PCV round.

Road surfaces

Roundabouts are junctions where considerable braking and acceleration take place. The road surface can become polished and be slippery, especially in wet weather.

Ensure that all braking and speed reduction is done in good time.

If you can see that it's clear to enter the roundabout, do so – provided you won't cause any traffic from your right to brake or swerve.

Cyclists and horse riders

It's often safest for cyclists and horse riders to take the outside path when turning right at large roundabouts. Watch for any signals and give them as much room as you safely can.

THE PCV DRIVING TEST

Mini-roundabouts

The same rules and procedures apply at mini-roundabouts as at full-scale roundabouts:

- you should give way to traffic approaching from the right
- because of the restricted space both entering and leaving these locations, it's essential to keep a constant check on the mirrors
- the rear of a long vehicle can easily 'clip' a car waiting to enter a mini-roundabout
- it's most unlikely that PCVs will be able to turn at a mini-roundabout without driving over the marked centre area
- you should position your vehicle so that it doesn't mount the kerb at the entrance or exit.

Double mini-roundabouts

These require even more care and planning, since traffic will often back up from one to the other at busy times. Make sure that there's room for you to move forward and that, by doing so, your vehicle won't block the whole system.

Although traffic is advised not to carry out U-turn manoeuvres at a mini-roundabout, be alert for any oncoming traffic doing so.

Avoid any signals that might confuse. Because of the limited space and the comparatively short amount of time that it takes to negotiate a mini-roundabout, it's important to give only signals that will help other road users.

If you have to drive over a raised mini-roundabout, do so slowly and carefully, so as not to damage your vehicle or cause discomfort to your passengers.

At any roundabout, cancel your indicator signal as soon as you've completed the manoeuvre.

Multiple roundabouts

At a number of (usually well-known) locations complex roundabout systems have been designed, which incorporate a mini-roundabout at each exit.

The main thing to remember at such places is that traffic is travelling in all directions.

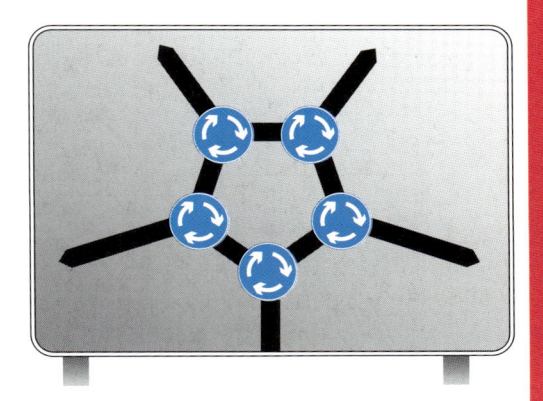

Sometimes mini-roundabouts are sited at what were formerly T-junctions. These junctions can be at a variety of angles, so you should adopt the safest position on approach (even if technically 'going ahead'). Give an appropriate signal to other road users.

213

THE PCV DRIVING TEST

Hazards – overtaking

What the test requires

When overtaking, you must

- look well ahead for any hazards, such as
  - oncoming traffic
  - bends
  - junctions
  - road markings
  - traffic signs
  - the vehicle in front about to overtake
  - any gradient
- assess the speed of the vehicle you intend to overtake
- assess the speed differential of the two vehicles. This will indicate how long the manoeuvre could take
- allow enough room to overtake safely
- avoid the need to 'cut in' on the vehicle you've just overtaken.

How your examiner will test you

For this aspect of driving there isn't a special exercise. The examiner will watch carefully and take account of your

- use of the MSM/PSL routine
- reactions to road and traffic conditions
- handling of the controls
- choice of safe opportunities to overtake.

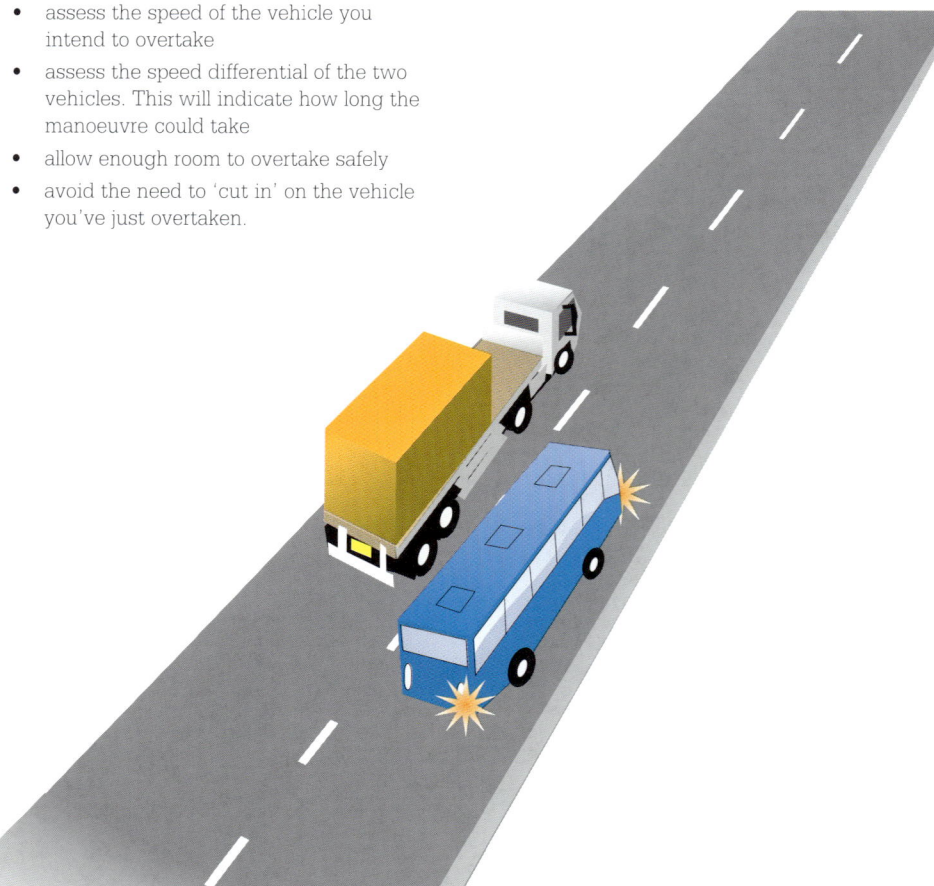

THE PCV DRIVING TEST

Skills you should show

You must be able to assess all the factors that will help you to decide if you can overtake safely, such as

- oncoming traffic
- the type of road (single or dual carriageway)
- the speed of the vehicle ahead
- if you can overtake before reaching any continuous white line on your side of the road
- how far ahead the road is clear
- whether the road will remain clear
- whether your mirror checks show that there's traffic behind about to overtake.

Overtake only when you can do so

- safely
- legally
- without causing other road users to slow down or alter course.

Faults to avoid

You must not overtake when

- your view of the road ahead isn't clear
- you would have to exceed the speed limit set for your vehicle on that type of road
- to do so would cause other road users to slow down, stop or swerve
- there are signs or road markings that prohibit overtaking.

215

THE PCV DRIVING TEST

Hazards – meeting and passing other vehicles

What the test requires

You must be able to meet and deal with oncoming traffic safely and confidently, especially

- on narrow roads
- where there are obstructions such as parked cars
- where you have to move into the path of oncoming vehicles.

How your examiner will test you

For this aspect of driving there isn't a special exercise. The examiner will watch carefully and take account of your

- use of the MSM/PSL routine
- reactions to road and traffic conditions
- handling of the controls.

Skills you should show

You should

- show sound judgement when meeting oncoming traffic

- be decisive when stopping and moving off
- stop in a position that allows you to move out smoothly when the way is clear
- allow adequate clearance when passing stationary vehicles. Slow down if you have to pass close to them.

Be on the alert for

- doors opening
- children running out
- pedestrians stepping out from between parked vehicles or round the front of other buses
- vehicles pulling out without warning.

Faults to avoid

You should avoid

- causing other vehicles to
    - slow down
    - swerve
    - stop
- passing dangerously close to parked vehicles
- using the size of your vehicle to force other road users to give way.

216

THE PCV DRIVING TEST

Hazards – crossing the path of other vehicles

What the test requires

You must be able to cross the path of oncoming traffic safely and with confidence. You'll need to be able to carry this out safely when you intend to

- turn right at a road junction
- enter bus stations or garages on the right-hand side of the road.

You should

- use the MSM/PSL routine on approach
- position the vehicle correctly. The width and type of road and the length of the vehicle will affect this
- accurately assess the speed of any approaching traffic
- wait, if necessary
- observe the road or entrance you're about to turn into
- watch for any pedestrians.

How your examiner will test you

For this aspect of driving there isn't a special exercise. The examiner will watch carefully and take account of your judgement of oncoming traffic.

Skills you should know

You should

- make safe and confident decisions about when to turn across the path of vehicles approaching from the opposite direction
- ensure that the road or entrance is clear for you to enter
- be confident that your vehicle won't endanger any road user waiting to emerge from the right
- accurately assess whether it's safe to enter the road or entrance
- show courtesy and consideration to other road users, especially pedestrians.

Faults to avoid

You should avoid

- turning across the path of oncoming road users, causing them to
  - slow down
  - swerve
  - brake
- cutting the corner so that you endanger vehicles waiting to emerge
- overshooting the turn so that the front wheels mount the kerb.

THE PCV DRIVING TEST

Hazards – pedestrian crossings

What the test requires

You must be able to

- recognise the different types of pedestrian crossing
- show courtesy and consideration towards pedestrians
- stop safely, when necessary.

How your examiner will test you

For this aspect of driving there isn't a special exercise. The examiner will watch carefully to see that you

- recognise the pedestrian crossing in good time
- use the MSM/PSL routine
- stop when necessary
- are especially alert when crossings are sited
  - near schools
  - in shopping areas
  - at or near junctions.

Skills you should show

You should

- approach all crossings at a controlled speed
- stop safely, when necessary
- move off when you're sure it's safe to do so.

Controlled crossings

These crossings may be controlled by

- traffic signals at junctions
- police officers
- traffic wardens
- school crossing patrols.

218

THE PCV DRIVING TEST

Zebra crossings

These crossings can be recognised by

- black and white stripes across the road
- a row of studs along each edge of the black and white stripes
- tactile paving on both sides of the crossing for partially sighted people
- flashing amber beacons at both sides of the road
- zigzag markings on the road on both sides of the crossing.

You must

- slow down and stop if there's anyone on the crossing
- slow down and be prepared to stop if anyone is waiting to cross or will reach the crossing before you do.

Pelican crossings

These crossings have

- traffic signals that change only after pedestrians have pressed a button on either side of the crossing
- a flashing amber phase to allow pedestrians already crossing to get across safely
- zigzag lines on the road on each side of the crossing
- a stop line painted on the road for traffic waiting at the crossing.

The sequence of the traffic lights is

- red
- flashing amber
- green
- amber
- red.

You must

- stop if the lights are on red or amber
- give way to any pedestrians crossing if the amber lights are flashing
- give way to any pedestrians still crossing when the flashing amber light changes to green.

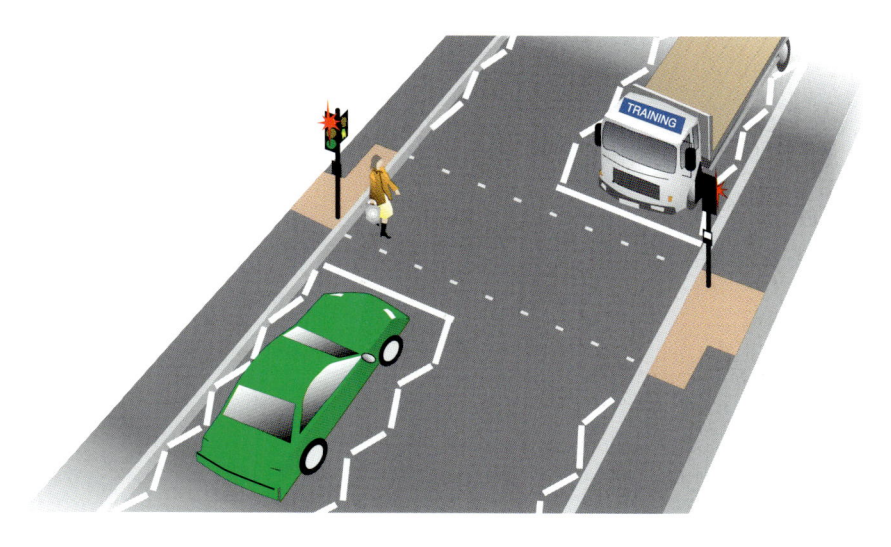

THE PCV DRIVING TEST

Puffin crossings

The term 'puffin' is an acronym for pedestrian user-friendly intelligent crossings. This type of crossing has been installed at a number of selected sites and can be identified by

- detectors sited so that the red traffic signal will be held until pedestrians have cleared the crossing
- no flashing amber phase
- traffic lights that operate in normal sequence
    - red
    - red and amber
    - green
    - amber
    - red.

You must

- stop and wait, unless the green light is showing
- drive over the crossing only if it's clear of pedestrians.

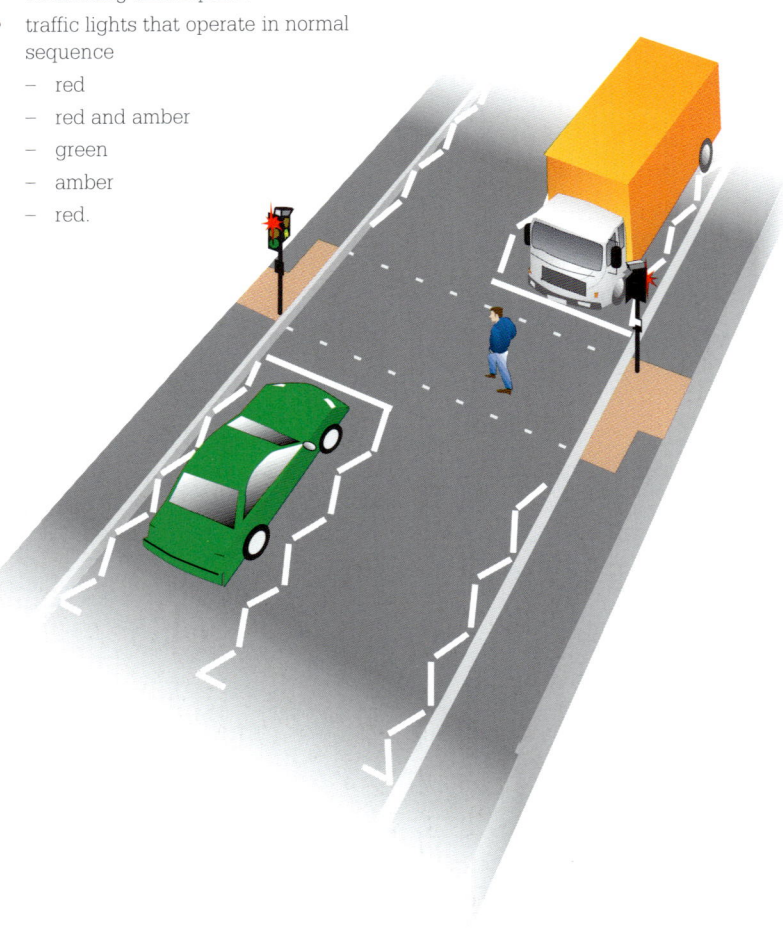

THE PCV DRIVING TEST

Toucan crossings

These crossings are mostly found in areas with college or university sites and where there are large numbers of cyclists. They operate in the same way as puffin crossings except

- cyclists share the crossing with pedestrians without dismounting
- a green cycle light indicates when it's safe to cross.

As with puffin crossings, the traffic lights operate in the normal sequence.

You must

- stop and wait, unless the green light shows
- drive over the crossing only if it's clear of pedestrians or cyclists.

Faults to avoid

You should avoid

- approaching any type of crossing at too high a speed
- driving on without stopping or showing awareness of waiting pedestrians
- driving onto or blocking a crossing
- harassing pedestrians by
  - revving the engine
  - making the air brakes hiss
  - edging forward
  - sounding the horn
  - overtaking within the zigzag lines
  - waving them to cross.

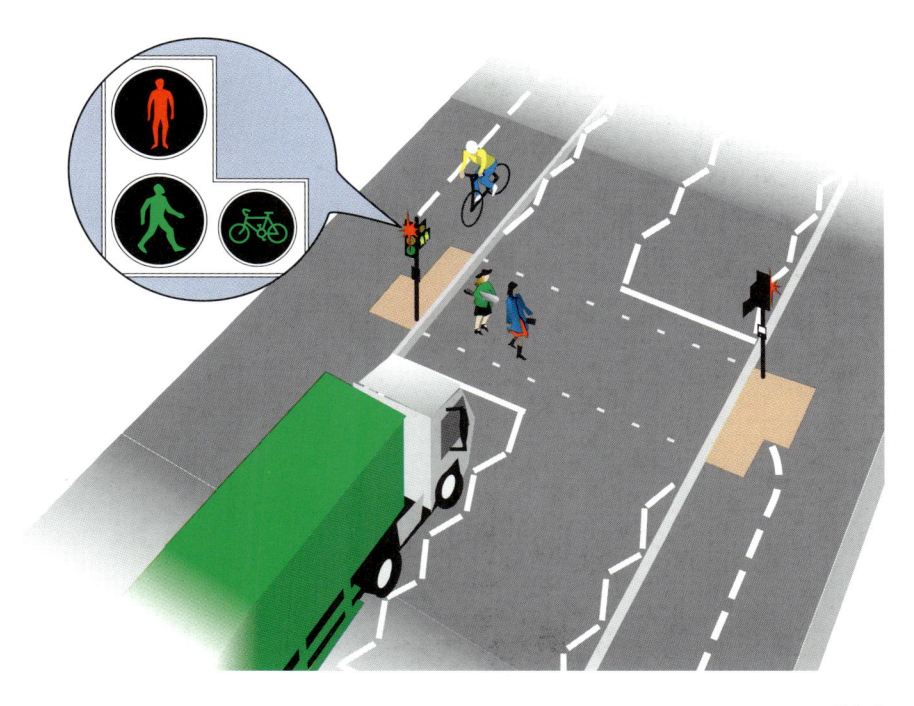

221

THE PCV DRIVING TEST

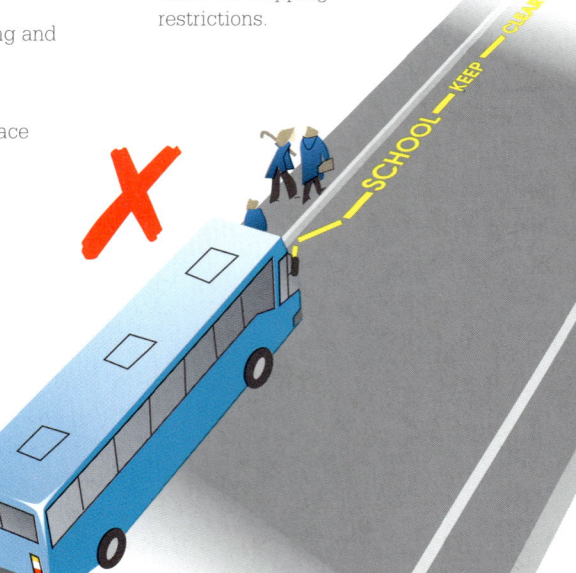

Selecting a safe place to stop

What the test requires

When you make a normal stop you must be able to

- select a safe place where you won't
  - cause an obstruction
  - create a hazard
  - contravene any waiting, stopping or parking restrictions
- stop reasonably close to the edge of the road.

How your examiner will test you

At times during the test the examiner will ask you to pull up either at

- a convenient place or
- a particular place, for example next to a lamp-post or, in some circumstances, at a bus stop.

This is to demonstrate that you could pull up to allow passengers to board or alight safely.

The examiner will watch your driving and take account of your

- use of the MSM/PSL routine
- judgement in selecting a safe place to stop.

Skills you should show

You must be able to stop in a safe position by

- selecting it in good time
- making proper use of the MSM/PSL routine
- only stopping where you're allowed to do so
- not causing an obstruction
- recognising in good time road markings or signs indicating any restriction
- pulling up close to and parallel with the kerb
- applying the parking brake while the vehicle is stationary
- stopping at the correct place when asked.

Faults to avoid

You should avoid

- pulling up with insufficient warning to other road users
- causing danger or inconvenience to any other road users
- parking at or outside
  - school entrances
  - fire stations
  - ambulance stations
  - pedestrian crossings.

You must comply with

- 'No Waiting' signs or markings
- 'No Parking' signs or markings
- other 'no stopping' restrictions.

THE PCV DRIVING TEST

Uncoupling and recoupling

What the test requires

If you're taking a test to gain a trailer entitlement, you'll be asked to uncouple and recouple your vehicle, normally at the end of the test. You should know and be able to demonstrate how to uncouple and recouple your vehicle safely.

Uncoupling

When uncoupling you should

- ensure that the brakes are applied on both the vehicle and trailer
- set the jockey wheel/prop stand to support the trailer weight
- turn off any taps, disconnect the air lines and stow the lines away safely (where fitted)
- disconnect the electric line and stow it away safely
- release the break-away cable connection
- release the trailer coupling
- drive the tractive unit away slowly, checking the trailer either directly or in the mirrors.

Recoupling

When recoupling

- ensure that the trailer brake is applied
- reverse slowly up to the trailer
- ensure that the vehicle parking brake is applied
- check the height of the coupling
- connect the tow-hitch
- connect the break-away cable
- connect the electric lines

- connect the air lines and turn on taps, if fitted
- raise the jockey wheel/prop stand
- release the trailer parking brake
- start up the engine
- check that the air is building up in the storage tanks (where applicable)
- check lights and indicators.

How your examiner will test you

Your examiner will ask you to perform this exercise where there's safe and level ground. You'll be asked to

- demonstrate the uncoupling of your vehicle and trailer
- pull forward and park the vehicle alongside the trailer
- realign the vehicle with the trailer before recoupling the trailer.

Your examiner will expect you to make sure that the

- coupling is secure
- lights and indicators are working
- the trailer brake is released.

223

THE PCV DRIVING TEST

Skills you should show

You should be able to uncouple and recouple your vehicle and trailer

- safely
- confidently, and in good time
- showing concern for your own and other's health and safety.

Faults to avoid

Uncoupling

When uncoupling you should avoid

- uncoupling without applying the brakes on the towing vehicle
- releasing the trailer coupling without the jockey wheel/ prop stand being lowered
- moving forward before the entire correct procedure has been completed.

Recoupling

When recoupling you should avoid

- not checking the brakes are applied on the trailer
- not using good, effective observation of your trailer as you reverse up to it
- leaving the towing vehicle without applying the parking brake
- recoupling at speed.

Don't attempt to move away without checking the

- lights
- indicators
- trailer brake release.

THE PCV DRIVING TEST

Understanding the rules

At the end of the test the examiner will ask you to show the

- location of the fire extinguisher
- fuel cut-off device
- emergency door and how it operates.

With the introduction of the theory test for large goods vehicle and PCV drivers, questions on *The Highway Code* won't be asked any longer at the end of the practical driving test. However, you'll be expected to

- put its rules into practice when you're driving
- recognise all road signs or road markings that apply to minibus, coach or bus drivers
- show courtesy and consideration towards all other road users.

The Highway Code itself isn't a set of laws, rather a collection of rules offering sound guidance to all road users. Know the rules and use them whenever you drive on the road.

New road signs are introduced from time to time, and the rules set out in *The Highway Code* may be amended or increased. You should ensure that you're familiar with the most recent edition.

You should also study and be totally familiar with all the signs and road markings set out in the book *Know Your Traffic Signs* (TSO). Changes to UK traffic signs will continue to take place over a number of years. It's your responsibility to be aware of any changes as they're introduced.

THE PCV DRIVING TEST

The test results

Legal requirements of the test

The candidate must show that they're competent to drive the vehicle in which the test is being conducted without danger to, and with due consideration for, other persons using the road. In particular, the candidate must show that they can competently

- start the engine
- move off straight ahead and at an angle
- maintain a proper position in relation to a vehicle immediately in front
- overtake and take an appropriate course in relation to other vehicles
- turn right and left
- stop within a limited distance, under full control
- stop normally and bring the vehicle to rest in an appropriate part of the road
- drive the vehicle forwards and backwards; whilst driving the vehicle backwards steer the vehicle along a predetermined course to make it enter a restricted opening and bring it to rest in a predetermined position
- indicate their intended actions by appropriate signals at appropriate times
- act correctly and promptly in response to all signals given by any traffic sign, by any person lawfully directing traffic, and by any other person using the road.

If you pass

You'll have demonstrated that you can drive a bus, coach or minibus – without passengers – to the high standard required to obtain a licence. You'll be given

- a pass certificate (D10V)
- a copy of the driving test report (DLV25A), which will show any driving faults that have been marked during the test

You'll also be offered a brief explanation of any driving faults marked. This is to help you overcome any weaknesses in your driving as you gain experience.

To apply for your full licence

- complete a D750 form to apply for a photocard licence
- enclose your licence
- enclose the pass certificate (D10V) with the applicant's declaration signed
- enclose the appropriate fee.

Send these to

The Vocational Section
DVLA
Swansea
SA99 1BR

as soon as you can (or in any case within two years).

After you've passed

You should aim to raise your standard of driving – especially as you'll be driving buses carrying passengers.

Most operators will offer you 'type' training, which will allow you to familiarise yourself with the different vehicles in the fleet.

Your trainer should be able to give you further advice.

THE PCV DRIVING TEST

If you don't pass

Your driving won't have been up to the high standard required to obtain the vocational driving licence. You'll have made mistakes which either caused, or could have caused, danger on the road.

Your examiner will

- give you a statement of failure including a copy of the driving test report (DLV25A), which will show all the faults marked during the test
- explain briefly why you've failed.

You should study the driving test report carefully and refer to the relevant sections in this book.

Show the report to your instructor, who will help you to correct the points of failure. Listen to the advice your instructor gives and try to get as much practice as you can before you retake your test.

Right of appeal

Although the examiner's decision can't be altered, you have a right to appeal if you consider that your driving test wasn't conducted according to the regulations.

If you live in England or Wales you have six months after the issue of the statement of failure in which to appeal (Magistrates' Courts Act 1952 [Ch. 55 part VII, Section 104]). If you live in Scotland you have 21 days in which to appeal (Sheriff Court, Scotland Act of Sederunt (Statutory Appeals) 1981).

See also the DSA complaints guide for test candidates at the back of this book.

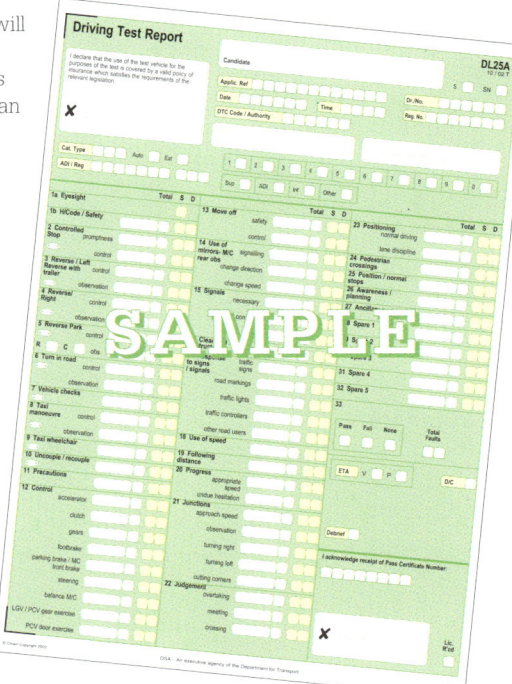

Part Seven
Additional information

The topics covered

- **Disqualified drivers**
- **DSA services**
- **Useful addresses**
- **PCV licence entitlements**
- **Minimum test vehicles**
- **New MTV requirements**
- **Vehicle types and licence requirements**
- **Cone positions**
- **Road signs**
- **Conclusion**
- **Glossary**

ADDITIONAL INFORMATION

Disqualified drivers

Retesting once disqualified

Tougher penalties now exist for anyone convicted of certain dangerous driving offences. If a driver is convicted of a dangerous driving offence, which involves a period of disqualification, all PCV entitlement is automatically lost regardless of the type of vehicle being driven at the time of the offence.

The decision about whether that entitlement can be regained is a matter for the Licensing Authority (LA). The options are

- the entitlement may be refused on the grounds that you've shown yourself to be an unfit and improper person to hold a bus or coach driving licence
- the court may require you to take an extended car driving test to regain your category B licence
- you may be required to retake a driving test for each additional category of vehicle that you want to drive
- the additional category(ies) may be restored without any further requirement, in exceptional circumstances.

It's important to remember that a PCV driving licence can't be issued on its own. You must possess a valid, full driving licence entitlement for category B (a car licence) for your category D, D1 or D + E licence entitlement to be valid. If you lose your car licence entitlement you lose your PCV licence with it.

ADDITIONAL INFORMATION

Applying for a retest

If you have to take a category B retest you can apply for a provisional licence at the end of the period of disqualification.

The normal rules for provisional licence-holders apply

- you must be supervised by a person who's at least 21 years of age and has held (and still holds) a full licence for at least three years for the category of vehicle being driven
- L plates (or D plates, if you wish, in Wales) must be displayed to the front and rear of the vehicle
- driving on motorways isn't allowed
- PCVs may not be driven if you've only a provisional car licence (category B).

All driving tests are booked by application to the DSA Booking Centre or by telephoning the National Booking number: 0870 01 01 372. There are higher fees for extended tests, so you must make it clear when you apply which type of test you want.

You can only apply for a provisional category D licence entitlement after you've passed an extended car driving test, if the court has directed you to do so.

ADDITIONAL INFORMATION

DSA services

Service standards

The Driving Standards Agency (DSA) is committed to providing a high-quality service for all its customers. If you would like information about our standards of service, please contact

Customer Service Unit
Driving Standards Agency
Stanley House
Talbot Street
Nottingham
NG1 5GU

Tel: 0115 901 2500/2545
Fax: 0115 901 2510
Email: customer.services@dsa.gsi.gov.uk

Complaints guide

DSA aims to give our customers the best possible service. Please tell us

- when we've done well
- when you aren't satisfied.

Your comments can help us to improve the service that we offer.

If you have any questions about how your test was conducted please contact the local Supervising Examiner, whose address is displayed in your local driving test centre. If you're dissatisfied with the reply or wish to comment on other matters you can write to the Area Manager (see the list of Area Offices at the back of this book).

If your concern relates to an Approved Driving Instructor you should write to

The Registrar of Approved Driving Instructors
Driving Standards Agency
Stanley House
Talbot Street
Nottingham
NG1 5GU

Alternatively, you may wish to write to

The Chief Executive
Driving Standards Agency
Stanley House
Talbot Street
Nottingham
NG1 5GU

If you remain dissatisfied, you can ask the Chief Executive to refer your complaint to the Independent Complaints Assessor. None of this removes your right to take your complaint to

- your Member of Parliament, who may decide to raise your case personally with the DSA Chief Executive, the Minister, or the Parliamentary Commissioner for Administration (the Ombudsman), whose name and address are given on page 237
- a magistrates' court (in Scotland to the Sheriff of your area) if you believe that your test wasn't conducted in accordance with the relevant regulations.

Before doing this you're advised to seek legal advice.

ADDITIONAL INFORMATION

Refunds of out-of-pocket expenses

DSA always aims to keep test appointments but occasionally tests have to be cancelled at short notice. DSA will normally refund the test fee, or re-book your test at no further charge, in the following circumstances

- where an appointment is cancelled by DSA – for whatever reason
- where an appointment is cancelled by the candidate, who gives at least ten clear working days' notice
- where the candidate keeps the test appointment, but the test doesn't take place or isn't completed for reasons not owing to the candidate or to any vehicle provided by him or her for the test.

In addition, DSA will normally consider reasonable claims from the candidate for financial loss or expenditure incurred by him or her, as a result of DSA cancelling a test at short notice (other than for reasons of bad weather). For example, a claim for the commercial hire of the test vehicle will normally be considered. Applications should be made to the Area Office where the test was booked.

This compensation code doesn't affect your existing legal rights.

ADDITIONAL INFORMATION

Useful addresses

DSA Test Enquiries and Booking Centre

DSA
PO Box 280
Newcastle-upon-Tyne
NE99 1FP

Tel: 0870 01 01 372
Welsh speakers: 0870 01 00 372
Minicom: 0870 01 07 372
Fax: 0870 01 02 372

DSA Head Office

Stanley House
56 Talbot Street
Nottingham
NG1 5GU

Tel: 0115 901 2500
Fax: 0115 901 2940

Traffic Commissioners and Traffic Area Offices

South-Eastern and Metropolitan London

Ivy House
3 Ivy Terrace
Eastbourne
BN21 4QT

Tel: 01323 452 400
Fax: 01323 726 679

Area covered:
Brighton and Hove
East Sussex
Greater London
Kent
Medway Towns
Surrey
West Sussex

North-Eastern & North-Western

Hillcrest House
386 Harehills Lane
Leeds
LS9 6NF

Tel: 0113 254 3290/1
Fax: 0113 248 9607

Area covered:
Blackburn with Darwen
Blackpool
Cheshire
Cumbria
Darlington
Derby City
Derbyshire
Durham
East Riding of Yorkshire
Greater Manchester
Halton
Hartlepool
Kingston upon Hull
Lancashire
Merseyside
Middlesbrough
North Lincolnshire
North East Lincolnshire
North Yorkshire
Northumberland
Nottingham
Nottinghamshire
Redcar & Cleveland
South Yorkshire
Stockton-on-Tees
Tyne & Wear
Warrington
West Yorkshire
York

Eastern

Terrington House
13–15 Hills Road
Cambridge
CB2 1NP

Tel: 01223 531 060
Fax: 01223 532 089

ADDITIONAL INFORMATION

Area covered:
Bedfordshire
Buckinghamshire
Cambridgeshire
Essex
Hertfordshire
Leicester
Leicestershire
Lincolnshire
Luton
Milton Keynes
Norfolk
Northamptonshire
Peterborough
Rutland
Southend-on-Sea
Suffolk
Thurrock

Wales and West Midlands

Cumberland House
200 Broad Street
Birmingham
B15 1TD

Wales

Tel: 0121 609 6835

West Midlands

Tel: 0121 609 6813

Fax for both areas

0121 608 1001

Area covered: Wales
All of Wales

Areas covered: West Midlands
Herefordshire
Shropshire
Staffordshire
Stoke-on-Trent & Telford
Warwickshire
West Midlands
Worcestershire
Wrekin

Scotland

J Floor
Argyll House
3 Lady Lawson Street
Edinburgh
EH3 9SE

Tel: 0131 200 4455
Fax: 0131 529 8501

Area covered:
All of Scotland and the Islands

Western

2 Rivergate
Temple Quay
Bristol
BS1 5DR

Tel: 0117 900 8577

Area covered:
Avon
Bath & North-East Somerset
Berkshire
Bournemouth
Bracknell Forest
Bristol
Cornwall
Devon
Dorset
Gloucestershire
Hampshire
Isle of Wight
North Somerset
Oxfordshire
Plymouth
Poole
Portsmouth
Reading
Slough
Somerset
Southampton
South Gloucestershire
Swindon
Torbay
West Berkshire
Wiltshire
Windsor & Maidenhead
Wokingham

ADDITIONAL INFORMATION

PCV driving test centres

London and the South-East
Canterbury
Croydon
Enfield
Gillingham
Guildford
Hastings
Lancing
Purfleet
Yeading

Midlands and Eastern
Alvaston (Derby)
Chelmsford
Culham
Featherstone (Wolverhampton)
Garrets Green (Birmingham)
Harlescott (Shrewsbury)
Ipswich
Leicester
Leighton Buzzard
Norwich
Peterborough
Swynnerton (Stoke-on-Trent)
Watnall (Nottingham)
Weedon (Northampton)

Northern
Berwick-on-Tweed
Beverley
Bredbury (Manchester)
Carlisle
Darlington
Gosforth (Newcastle)
Grimsby
Kirkham (Preston)
Patrick Green (Leeds)
Sheffield
Simonswood
Steeton (Keighley)
Walton (York)

Scotland
Aberdeen
Benbecula*
Bishopbriggs (Glasgow)
Dumfries
Elgin
Galashiels
Inverness
Kilmarnock
Kirkwall
Lerwick
Livingstone (Edinburgh)
Machrihanish (Kintyre)*
Perth
Port Ellen (Islay)*
Portree*
Stornoway*
Wick

Wales and Western
Bristol
Caernarfon*
Camborne
Chisledon (Swindon)
Exeter
Gloucester
Haverfordwest (Withybush)
Llantrisant
Neath
Plymouth
Pontypool
Poole
Reading
Rookley (Isle of Wight)
Southampton
Taunton
Wrexham

*Tests are only conducted occasionally at these centres

ADDITIONAL INFORMATION

Bus and Coach Working Group (DPTAC)

The DETR Mobility Unit
111 Great Minster House
76 Marsham Street
London
SW1P 4DR

Tel: 020 7890 4916

City and Guilds of London Institute

1 Giltspur Street
London
EC1A 9DD

Tel: 020 7294 2468

Community Transport Association

Highbank
Halton Street
Hyde
Cheshire
SK14 2NY

Tel: 0161 366 6685
Fax: 0161 351 7221

Confederation of Passenger Transport UK

(previously the Bus and Coach Council)

Imperial House
15–19 Kingsway
London
WC2B 6UN

Tel: 020 7240 3131
Fax: 020 7240 6565
Website: www.cpt-uk.org/cpt

Department for Transport (DfT)

Great Minster House,
76 Marsham Street,
London
SW1P 4DR

Tel: 020 7944 3000

Disabled Persons' Transport Advisory Committee (DPTAC)

Zone 1–4
Great Minster House
76 Marsham Street
London
SW1 4DR

Tel: 020 7944 3632
Minicom: 020 7944 3277
Fax: 020 7944 6998

Driver and Vehicle Licensing Agency (DVLA) Customer Enquiry Unit

Licence Enquiries
Swansea
SA6 7JL

Tel: 0870 240 0009
Minicom: 01792 782 787
Fax: 01792 783 071

(Ring between 8.15 am and 8.30 pm Monday to Friday, and between 8.30 am and 5.00 pm on Saturdays)

DfT Mobility Advice and Vehicle Information Service (MAVIS)

'O' Wing, Macadam Avenue
Old Wokingham Road
Crowthorne
RG45 6XD

Tel: 01344 661 000
Fax: 01344 661 066

DVLA Drivers' Medical Group

Swansea
SA99 1TU

Tel: 0870 600 0301

ADDITIONAL INFORMATION

GMPTE

9 Portland Street
Piccadilly Gardens
Manchester
M60 1HX

Tel: 0161 242 6243 (Minicom facility)
Fax: 0161 242 6139

Historic Commercial Vehicle Society

Iden Grange
Cranbrook Road
Staplehurst
Kent
TN12 0ET

Tel: 01580 892 929
Fax: 01580 893 227
Email: hcvs@btinternet.com
Website: www.hcvs.co.uk

HSE Infoline

HSE Information Services
Caerphilly Business Park
Caerphilly
CF83 3GG

Tel: 0870 154 5500
Fax: 02920 859 260

(See your telephone book for details of your area HSE office.)

London Transport Users' Committee

Clements House
14/18 Gresham Street
London
EC2V 7PR

Tel: 020 7505 9000
Fax: 020 7505 9003

Metropolitan Police Traffic Department

Traffic Headquarters
Room 1130
New Scotland Yard
Broadway
London
SW1H 0BG

Tel: 020 7230 3591

National Federation of Bus Users

PO Box 320
Portsmouth
PO5 3SD

Tel: 023 9281 4493

National Playbus Association

93 Whitby Road
Bristol
BS4 4AR

Tel: 0117 977 5375

Parliamentary Commissioner for Administration (The Ombudsman)

Ann Abraham
Millbank Tower
Millbank
London
SW1P 4QP

Tel: 0845 015 4033

Road Operator's Safety Council

395 Cowley Road
Oxford
OX4 2DJ

Tel: 01865 775 552
Fax: 01865 711 745

ADDITIONAL INFORMATION

Royal Society for the Prevention of Accidents (RoSPA)

Edgbaston Park
353 Bristol Road
Birmingham
B5 7ST

Tel: 0121 248 2000

RTITB

Ercall House
8 Pearson Road
Central Park
Telford
TF2 9TX

Tel: 01952 520 200
Fax: 01952 520 201

Transfed Ltd

Regency House
43 High Street
Rickmansworth
Hertfordshire
WD3 1ET

Tel: 01923 896 607
Website: www.transfed.org

Vehicle and Operator Services Agency (VOSA)

(formerly Transport Area Network and The Vehicle Inspectorate)

The Enquiry Unit
Welcombe House
91-92 The Strand
Swansea
SA1 2DA

Tel: 0117 954 3200
Fax: 01792 454 313

ADDITIONAL INFORMATION

PCV licence entitlements

The licence entitlements you'll require to drive different types of buses, coaches and minibuses are listed here. You must hold full (not provisional) category B entitlement before you can take a test in this group. You must also gain a full category entitlement for a vehicle before taking a second test to add the trailer entitlement (+ E). No additional entitlement is required to tow trailers that weigh less than 750 kg.

Category	Description	Additional categories covered
D	Any bus including articulated (or 'bendi-bus') with more than 8 passenger seats	D1
D1	Buses with 9–16 passenger seats	None
D + E	Buses towing trailers over 750 kg	D, D1, D1 + E
D1 + E	Buses with 9–16 passenger seats towing trailers over 750 kg, provided the combination doesn't exceed 12 tonnes and the laden trailer weight doesn't exceed the unladen weight of the towing vehicle	D1

If the vehicle you use for your driving test has automatic transmission, your licence entitlement won't include vehicles with manual gearboxes. A vehicle with automatic transmission is defined as a vehicle in which the driver isn't provided with any means whereby he or she may, independently of the use of the accelerator or the brakes, vary the proportion of the power being produced by the engine that's transmitted to the road wheels of the vehicle. This definition includes semi-automatic vehicles, where no clutch pedal exists.

Minibuses may only be driven with a category B licence entitlement within the UK provided

- the vehicle is used by a non-commercial body for social purposes only
- the driver is 21 years or more and has held a full car licence for at least two years
- the driver provides his or her services on a voluntary (unpaid) basis
- the minibus weighs no more than 3.5 tonnes (or 4.25 tonnes if specially adapted for disabled passengers).

ADDITIONAL INFORMATION

Minimum test vehicles

All test vehicles in this group must be unladen and capable of 80 kph (50 mph)

Category	Description
D	Any PCV with more than 8 passenger seats and at least 9 metres (29 feet 3 inches) long
D1	Any PCV with 9–16 passenger seats*
+ E (Towing trailers)	If you want an entitlement to tow a trailer over 750 kg behind a bus in one of the above categories, you must tow an unladen trailer with a maximum authorised mass (MAM) of at least 1,250 kg behind the vehicle you use throughout the driving test. You must have passed a driving test using such a vehicle without a trailer before you can take this additional test. If you already have a C + E entitlement this requirement is waived.

***Important note** A vehicle with more than 17 seats, including the driver's, isn't suitable for category D1. A vehicle must fall within the category or sub-category for which the licence is being sought **in addition to meeting the MTV specification**.

PCVs with restricted rear vision such as Highliners and Neoplans are suitable vehicles for test, providing the braking manoeuvre is carried out off-road.

Stretched limousines and prison vans based on a lorry chassis are not suitable vehicles for a PCV test.

A vehicle carrying a trade plate is not suitable for a driving test, as the conditions attached to trade licences do not allow for a vehicle to be used for this purpose.

ADDITIONAL INFORMATION

New regulations

In September 2000, the European Commission amended the minimum driving test requirements that must be adopted by all member states so that testing arrangements are kept relevant to modern driving and riding conditions.

Changes to driving test arrangements include more demanding minimum test vehicle requirements, with larger and heavier vehicles for lorry, bus and vehicle-trailer tests. The additional requirements are shown below.

New MTV requirements for vehicles used for driving tests will apply to all vehicles brought into first use (first registration) from 1 October 2003. Vehicles brought into first use (first registration) before 1 October 2003 that meet the current MTV standards can be used for driving tests until the end of June 2007.

Further information and advice can be obtained on DSA's website: www.driving-tests.co.uk, or by telephoning DSA's Policy Section on: 0115 901 2569.

New additional MTV requirements

Category	Description
D	Minimum length 10 metres, minimum width 2.4 metres
D1	Minimum length 5 metres; maximum authorised mass 4 tonnes
D + E	Minimum width 2.4 metres; maximum authorised mass of trailer 1.25 tonnes; trailer cargo compartment must be closed box body at least 2 metres wide and 2 metres high
D1 + E	Maximum authorised mass 4 tonnes; maximum authorised mass of trailer 1.25 tonnes; trailer cargo compartment must be closed box body at least 2 metres wide and 2 metres high

Category D and D1 test vehicles must be fitted with an anti-lock brake system and a tachograph.

ADDITIONAL INFORMATION

Vehicle types and licence requirements

You should check with DVLA if you're in any doubt as to the licence entitlement that you require. The entitlement you'll need for the types of vehicles listed in this book are as follows

Type of vehicle	Category required	Notes
People carrier/small minibus with fewer than 9 passenger seats	B	May be subject to taxi or private hire vehicle regulations if used commercially
Minibus or midibus with more than 8 and not more than 16 passenger seats	D1	D1 allows passengers to be carried for hire or reward
Single-deck service bus or midibus with more than 16 passenger seats	D	
Coaches with more than 16 passenger seats	D	
Buses towing trailers over 750 kg	D+E	
'Supertrams'	B	Further qualifications are required to comply with the light rail transit (LRT) systems regulations
Double-deck service buses and coaches (including those with 3 or 4 axles)	D	
Historic buses and coaches, i.e. vehicles over 30 years old	D	In some cases these may be driven with category B entitlement – when not being used for hire or reward, or the carriage of more than 8 passengers
Mobile project and playbuses	C	In some cases these may be driven with category B entitlement
Towing trailers	+ E	In addition to the vehicle category

ADDITIONAL INFORMATION

Cone positions

Ready reckoner: metric measurements

This list of metric measurements should prove useful if you want to practise the reversing exercise.

To calculate the reversing area's layout identify the length of your vehicle in the left-hand columns and scan across to the right-hand columns for the relevant cone measurements. The cone positions are relative to the base line Z (see diagram on page 176).

Metres	Feet	Cone A	Cone B
4.50	14.8	22.5	13.5
4.75	15.6	23.8	14.3
5.00	16.4	25.0	15.0
5.25	17.2	26.3	15.8
5.50	18.0	27.5	16.5
5.75	18.9	28.8	17.3
6.00	19.7	30.0	18.0
6.25	20.5	31.3	18.8
6.50	21.3	32.5	19.5
6.75	22.1	33.8	20.3
7.00	23.0	35.0	21.0
7.25	23.8	36.3	21.8
7.50	24.6	37.5	22.5
7.75	25.4	38.8	23.3
8.00	26.2	40.0	24.0
8.25	27.1	41.3	24.8
8.50	27.9	42.5	25.5
8.75	28.7	43.8	26.3
9.00	29.5	45.0	27.0
9.25	30.3	46.3	27.8
9.50	31.2	47.5	28.5
9.75	32.0	48.8	29.3
10.00	32.8	50.0	30.0
10.25	33.6	51.3	30.8
10.50	34.4	52.5	31.5
10.75	35.3	53.8	32.3
11.00	36.1	55.0	33.0
11.25	36.9	56.3	33.8
11.50	37.7	57.5	34.5
11.75	38.5	58.8	35.3
12.00	39.4	60.0	36.0
12.25	40.2	61.3	36.8
12.50	41.0	62.5	37.5
12.75	41.8	63.8	38.3
13.00	42.7	65.0	39.0
13.25	43.5	66.3	39.8
13.50	44.3	67.5	40.5
13.75	45.1	68.8	41.3
14.00	45.9	70.0	42.0
14.25	46.8	71.3	42.8
14.50	47.6	72.5	43.5
14.75	48.4	73.8	44.3
15.00	49.2	75.0	45.0
15.25	50.0	76.3	45.8
15.50	50.9	77.5	46.5
15.75	51.7	78.8	47.3
16.00	52.5	80.0	48.0
16.25	53.3	81.3	48.8
16.50	54.1	82.5	49.5
16.75	55.0	83.8	50.3
17.00	55.8	85.0	51.0
17.25	56.6	86.3	51.8
17.50	57.4	87.5	52.5
17.75	58.2	88.8	53.3
18.00	59.1	90.0	54.0
18.25	59.9	91.3	54.8

ADDITIONAL INFORMATION

Road signs

You must be aware of the specific road signs that relate to buses and coaches. Those illustrated on this page are currently in use.

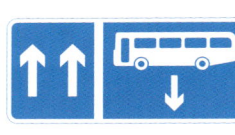

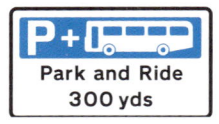

ADDITIONAL INFORMATION

Conclusion

Buses, coaches, minibuses and trams have developed rapidly over the past decade or so. Modern vehicles are fitted with 'smart engines' and 'thinking gearboxes', and the driver is surrounded by all manner of electronic circuitry to make the job easier, less stressful and often very enjoyable. The manufacturers have listened to the needs of drivers, passengers and operators equally when designing their products.

Today's bus driver should have 'service to the customer' as a primary aim. However, to give a professional service you need to be a skilled, dedicated driver – your driving should be to the highest standards. Your vehicle has made that easier, but the responsibilities you have are greater than they've ever been.

It's in your own interest to keep up to date with changes in requirements as they occur. Ignorance is no defence in law. Read the informative articles that appear in magazines devoted to driving. Remember, by passing the PCV driving test you'll only just be setting out on your career. Driving is a life skill that needs constant practice and revision.

You may drive a vehicle that doesn't require you to take a test, or have taken your test some time ago: by reading this book, you've shown that you understand the need for high standards in your driving. If you're a 'bus enthusiast' you'll find that you enjoy your interest more as your depth of knowledge and understanding increases. Whatever your reason, by studying this book you'll have made your objective

safe driving for life.

ADDITIONAL INFORMATION

Glossary

A *ABS* Anti-lock braking system (developed by Bosch) that uses electronic sensors to detect when a wheel is about to lock, releases the brakes sufficiently to allow the wheel to revolve, then repeats the process in a very short space of time – thus reducing the risk of skidding.

AETR rules European agreement concerning the work of crews on vehicles engaged in international road transport (aligned with EU regulations in April 1992). These rules govern drivers' hours and rest periods in specified countries outside the EU. For more detailed information please consult PSV375, published by the Department for Transport, Local Government and the Regions.

Air suspension system This uses a compressible material (usually air), contained in chambers located between the axle and the vehicle body, to replace normal steel-leaf spring suspension. Gives even height (empty or laden) and added comfort to passengers. It's also known as 'road-friendly' suspension.

Axle weights Limits laid down for maximum permitted weights carried by each axle.

C *C & U (Regs)* Construction and Use regulations that govern the design and use of all vehicles.

CAG Computer-aided gearshift system, developed by Scania, that employs an electronic control unit combined with electro-pneumatic actuators and a mechanical gearbox. The clutch is still required to achieve the gear change using an electrical gear lever switch.

COSHH Regulations 1988 The Control of Substances Hazardous to Health Regulations 1988 place a responsibility on employers to make a proper assessment of the effects of the storage or use of any substances that may represent a risk to their employees' health. Details can be obtained from the Health and Safety Executive.

CPC Certificate of Professional Competence indicates that the holder has attained the standards of knowledge required in order to exercise proper control of a transport business (required before an operator's licence can be granted).

Cruise control A facility that allows a vehicle to travel at a set speed without use of the accelerator pedal. However, the driver can immediately return to normal control by pressing the accelerator or brake pedal. This is rarely fitted to large PCVs, but may be found on some minibuses.

D *DPTAC* (Disabled Persons' Transport Advisory Committee) specification Applied to PCVs to assist passengers with disabilities (for example, bright yellow handrails, etc.).

Double de-clutching A technique employed when driving older PCVs that allows the driver to adjust the engine revs to the road speed when changing gear. The clutch pedal is released briefly while the gear lever is in the neutral position. When changing down, engine revs are increased to match the engine speed to the lower gear in order to minimise the load being placed on the gear mechanism.
Note: the construction of modern synchromesh gearboxes is such that this technique can cause damage. At least one major manufacturer has made it clear that the warranty conditions will become invalid if this technique has been used. Refer to the manufacturer and the vehicle handbook, if in doubt.

ADDITIONAL INFORMATION

Drive-by-wire Modern electronic and air control systems that replace direct mechanical linkages.

E *Electronic engine management system* This system monitors and controls both fuel supply to the engine and the contents of the exhaust gases produced. The system is an essential part of some speed-retarder systems.

Electronic power shift A semi-automatic transmission system that requires the clutch to be fully depressed each time a gear change is made. The system then selects the appropriate gear.

Endurance braking See *Retarder*.

F *Fluid flywheel* Incorporated in automatic and semi-automatic gear systems, it couples the drive train to the gearbox by use of hydraulic fluid. This allows gear changes, stopping and starting without the need for a separate clutch.

G *Geartronic* A fully automated transmission system developed by Volvo There's no clutch pedal, but an additional pedal operating an exhaust brake instead.

GVW Gross vehicle weight, applying to vehicles that include fuel, passengers, etc.

H *HSE* The Health and Safety Executive. HSE produces literature that provides advice and information on health and safety issues at work.

I *ISO 9000* British Standards code relating to quality assurance adopted by vehicle body-builders, recovery firms, etc.

J *Jake brake* A long-established system of speed retarding that alters the valve timing in the engine. In effect, the engine becomes a compressor and holds back the vehicle's speed.

K *Kerb weight* The total weight of a vehicle plus fuel, excluding any load (or driver).

'Kneeling' bus This type of bus uses air suspension to lower the entrance of the bus, whilst stationary, for easier access – especially useful for disabled passengers.

L *Lamination* A process whereby plastic film is sandwiched between two layers of glass so that an object, upon striking a windscreen, for example, will normally chip or craze the screen without large fragments of glass causing injury to the driver.

Limited-stop service A bus service operating under stage carriage conditions, but stopping only at specified points.

LNG Liquefied (compressed) Natural Gas.

Load-sensing valve A valve in an air brake system that can be adjusted to reduce the possibility of wheels locking when the vehicle is unladen.

LPG Liquefied (compressed) Petroleum Gas.

P *Plate* A plate fixed to the vehicle with information relating to dimensions and weights of passenger vehicles. It indicates tyre size, maximum axle weight and maximum loaded weight. Certificates are sometimes referred to as 'plates' when required, with information relating to tachographs, speed limiters, manufacturer's specification and height.

Pneumo-cyclic gearbox A semi-automatic gearbox, where an electronic or mechanical gearshift operates an air valve system to change gears.

ADDITIONAL INFORMATION

Pre-selector gearbox A gear-change system where the gears are manually selected prior to use and then engaged by pressing a gear-change pedal. It employs a fluid flywheel and no clutch.

[R] *Range change* A gearbox arrangement that permits the driver to select a series of either high or low ratio gears depending on the load, speed and any gradient being negotiated. Rarely fitted to PCVs.

Red Routes Approximately 300 route miles in the London area are subject to stringent regulations restricting stopping, unloading and loading.

Re-grooving A process permitted for use on tyres for vehicles over an unladen weight of 2,540 kg, allowing a new tread pattern to be cut into the existing tyre surface (subject to certain conditions).

Retarder or endurance brake An additional braking system that may be either mechanical, electrical or hydraulic. Mechanical devices either alter the engine exhaust gas flow or amend the valve timing (creating a 'compressor' effect). Electrical devices comprise an electromagnetic field energised around the transmission drive shaft (most frequently used on passenger vehicles). This type may also be known as 'regenerative' braking, when the energy generated is fed back into the vehicle's electrical storage system (batteries).

[S] *Semi-automatic* A transmission system in which there's no clutch but the driver changes gear manually.

Skip change Also known as block change, this sequence of gear-changing omits intermediate gears. Sometimes referred to as 'selective' gear-changing.

Splitter box Another name for a gearbox with high and low ratios.

[T] *Tachograph* A recorder indicating vehicle speeds, duration of journey, rest stops, etc. Required to be fitted to specified vehicles.

TBV Initials of the French (Renault) semi-automatic transmission system that employs a selector lever plus visual display information.

TC Traffic Commissioner appointed by the Secretary of State to a Traffic Area to act as the licensing authority for Goods Vehicle and PSV operators in the area.

Thinking gearbox A term used to describe a fully automated gearbox that selects the appropriate gear for the load, gradient and speed, etc. by means of electronic sensors.

Toughened safety glass The glass undergoes a heat treatment process during manufacture so that in the event of an impact on the windscreen (a stone, etc.) it breaks up into small blunt fragments, thus reducing the risk of injury. An area in front of the driver is designed to give a zone of vision in the event of an impact. The majority of bus and coach windscreens are laminated glass but toughened glass is used for other windows in a bus or coach.

Turbo-charged The exhaust gas drives a turbine, which compresses incoming air and effectively delivers more air to the engine than is the case with a normal or non-turbo-charged engine.

Turbo-cooled Refers to a system where the air from the turbo-charger is cooled before being delivered to the engine. The cooling increases the density of the compressed air to further improve engine power and torque.

ADDITIONAL INFORMATION

Two-speed axle An electrical switch actuates a mechanism in the rear axle that doubles the number of ratios available to the driver.

U *Unloader valve* A device fitted to air brake systems, between the compressor and the storage reservoir, pre-set to operate when sufficient pressure is achieved and allowing the excess air to be released (often heard at regular intervals when the engine is running).

V *VED* Vehicle excise duty or the road fund licence.

INDEX

ABS (anti-lock braking systems) 52-3, 246
Acceleration 25, 28, 183
   sudden 25, 26, 27, 64, 123
Accelerator skills 183
Accidents 11, 60-1, 74, 117, 132-41, 167
   involving dangerous goods 135, 136, 137
   motorcyclists 133, 139
   passenger care at 134
   reporting 61, 133, 134, 135-6
   at the scene of 133-6
Aggression 12, 90-1
Air auxiliary systems 55, 164
Air brake systems 14, 53, 54
Air reservoirs 54
Air suspension 38, 66, 86, 246
   levelling systems 30
Alarm systems, security 67, 87
Alcohol 9, 75
Animals 205
Anti-lock braking systems (ABS) 52-3, 246
Anti-theft measures 67, 87
Anticipation 64, 89, 90, 203
Articulated buses 36, 63, 239
Asleep at the wheel, falling 74, 100, 101, 107
Attitude 12-16
Audible warning systems 67
Audio equipment 110
Automatic transmission 5, 35, 49, 65, 185-6
Auxiliary air systems 55, 164
Auxiliary lighting 15, 76, 102-3
Awareness 89, 96, 203

Bell codes 42
Belts, seat 32, 34, 35, 169
'Bendi-buses' 6, 36, 63, 239
Bleepers, reversing 67
Blind passengers 17, 20, 21
Blind spots 92, 96, 112

Blow-outs 27, 39, 66, 93, 108, 131
Books to study 3, 18, 46, 68, 78, 131, 149
Boredom 74, 100
Brake (*see also* Brakes, Braking)
   fade 28, 53
   hand 188, 189
   lights 96, 113
   parking 52, 53
   pressure warning devices 54
   secondary 52
   service 52, 53
Brakes 52-4 (*see also* Brake, Braking)
   emergency 189
   use of, for test 187-9
Braking (*see also* Brake, Brakes)
   controlled 28, 29
   correctly 25
   distance 116
   effort 27
   exercise for test 180-1
   in good time 29, 64, 123
   loss of control 26
   sudden 25, 26, 27, 28, 96-7, 123
Braking systems 52-4 (*see also* Brake, Brakes)
   air 14, 53, 54
   anti-lock (ABS) 52-3, 246
   controls 54
   endurance ('retarders') 53, 64, 124, 159, 248
   inspection 54, 122
   maintenance 54
   safety 53
Breakdowns 129-30, 167
   on motorways 118-19
   passenger safety 119, 120, 121, 130, 134
Bridges, collision with 60-1
Buffeting 16
Built-up areas at night 103
Bus lanes 83
Buses
   articulated 36, 63, 239

INDEX

double-deck service 35
historic 41
'kneeling' 22, 34, 247
mobile project 40
play 40
single-deck service 34
tri-axle 39

Cameras 79, 80
Cats' eyes 127
Centre of gravity 27
Centrifugal force 27
Charts, tachograph 68-9
Checks, daily 46, 77 (see also Maintenance)
Children 90, 94, 216
 carrying 18-19, 21, 32, 34, 169
Clearances (see Limits)
Clearways 82
Clutch 49
Clutch control skills 26, 183
Coaches
 double-deck 38
 historic 41
 single-deck 37
 tri-axle 39
Coasting 184
Cockpit drill 77
Commercial pressure 19
Compensation code 232
Competition 19, 91
Complaints guide 231
Concentration 74, 101, 107, 118
Congested roads 16
Congestion charging 82-3
Consideration 19, 20, 22
Contact lenses 7
Contraflows 118
Control, maintaining 25, 29, 89
Controls, understanding, for test 159-60, 182-92
Convoys 91
Coupling system 52

Courtesy 19, 20, 21, 22
Crosswinds 115, 128
Customer care 17
Cyclists 12, 16, 92, 93, 94, 103, 132, 203, 204, 212

D plates 158, 172, 230
D1 Form 5
D4 Form 5, 8
DLV25 driving test report form 147
DLV25A driving test report 226, 227
DLV26 test application form 152, 153
DSA (Driving Standards Agency) iii, 231-3
DVLA (Driver and Vehicle Licensing Agency) 5, 236
Daily checks 46, 77 (see also Maintenance)
Dangerous goods, accidents involving 135, 136, 137
Deaf passengers 17, 20, 21
Defects 11, 19, 76, 77, 86, 130 (see also Maintenance)
Delays 19
Diesel engines 46-7, 66
Diesel fuel system 47
Diesel spillages 66
Disability 151
Disability Discrimination Act 30
Disabled passengers 20-2, 30
Disqualified drivers 153
 retesting 155, 229-30
Double-deck coaches 38
Double-deck service buses 35
'Double de-clutching' 41, 246
Drinking and driving 9, 75
Drivers' hours of work 68-72
 domestic 71
 EC rules 68, 70-1
 events and emergencies 72
 mixed EC and domestic driving 72
 regular services 71
Drivers Medical Group 7, 236

251

INDEX

Driving
  forces (*see* Forces at work)
  at night 100-6
    in built-up areas 103
    hazards 102-3
    problems encountered 100
    in rural areas 104
  regulations 78-83
  in shopping areas 16, 204
  skills 89
  skills when coming across other road users 90, 92
  test (*see* Test, PCV driving)
Driving Standards Agency iii, 231-3
Drugs 9, 75, 107
Dual carriageways 208

EC drivers' hours rules 68, 70-1, 72
Elderly people 22, 90, 94
Electric shock 141
Electrical system 50
Emergency vehicles 117, 134
Endurance braking systems ('retarders') 53, 64, 124, 159, 248
Engine, construction and functioning of the internal combustion 46-7
Engine coolant 48, 110, 123
Engine lubrication system 47-8
Environmental issues 64-7, 145
Exhaust emissions 64, 66-7
Eye contact 17, 21
Eyesight test 7, 101

Falling asleep at the wheel 74, 100, 101, 107
Fire 99, 136-7
  extinguishers 137
First aid 133, 138-41
  bleeding 140
  burns 140
  equipment 138
  motorcyclists 133, 139
  resuscitation 139
  shock 133, 140
    electric 141
  training 138
  unconscious victims 139
Fitness 8, 107 (*see also* Health)
Flashing headlights 15, 16, 97
Fog 117, 126
  lights 70, 102
    high intensity rear 102, 109, 113, 117, 123, 126
  at night 104
Force, centrifugal 27
Forces at work 25-8
  gravity 27
  inertia 28
  kinetic energy 28, 64
  momentum 28
Forms
  D1: 5
  D4: 5, 8
  DLV25: 147
  DLV25A 226, 227
  DLV26: 152, 153
Friction 26
Frost 116
Fuel 65, 66-7, 110, 122
Fuel economy 64, 96
Fuel system, diesel 47
Fuels, alternative 65, 66-7

Gear changing 26, 29, 41, 191
Gearboxes 5, 34, 35, 48, 49, 172, 184-6
Glasses 7
Gravity 27
Gritting vehicles 125

Handbrake 188, 189
Hazard warning lights 96, 99, 113, 119, 131, 133
Hazards 203-20
  animals 205
  crossing other traffic 217
  cyclists 203, 204, 212

INDEX

horse riders 205, 212
junctions 208-9
lane discipline 206-7
meeting other vehicles 216
motorcyclists 203, 204
at night 102-3
other road users 203-4
overtaking 214-15
passing other vehicles 216
pedestrian crossings 218-21
pedestrians 203, 204
positioning 206-7
road surfaces 212
roundabouts 210-13 (*see also* Roundabouts)
  mini and multiple 213
shopping areas 16, 204
turning left or right 211-12, 217
Headlights 15, 99, 103, 104, 105, 113, 116, 127
  flashing 15, 16, 97
Health 8, 9, 74, 107
Health and safety 86-7
Hearing-impaired passengers 17, 20, 21
Height 59, 60
  guide 61
  limits 59
  notices 59
Highway Code 3, 10, 15, 97, 104, 149, 198, 225
Historic vehicles 41
Horn 15, 16, 67, 98
Horse riders 16, 90, 94, 205, 212
Hours of work (*see* Drivers' hours of work)

Icy weather 113, 116, 123
Illness 8, 9, 74, 107
Indicator lights 16, 97, 109
Inertia 28
Instruments 109, 192
Interior lights 102
Intimidation 14, 16

Junctions 76, 93, 94, 208-9

Kerb, distance from 16, 17
Kerb, running over 27, 35, 66, 93
Kinetic energy 28, 64
'Kneeling' buses 22, 34, 247

L plates 158, 172, 230
LRT systems 43-4
Lane discipline 112, 206-7, 210
Learner, driver accompanying 6
Learning disabilities, passengers with 22
Legal requirements 157-8, 173, 226
Length limits 63
Licence, applying for 5
Licence, provisional 5, 6, 151, 157, 171
Licence requirements 242
Licences, categories of 5, 6, 33, 36, 40, 41, 171, 239, 242
Lifts, passenger 22, 33
Light rail transit systems 43-4
Lighting up times 101
Lights 100, 101, 102-3 (*see also* Headlights)
  auxiliary 15, 76, 102-3
  brake 96, 113
  in fog 102, 109, 113, 117, 123, 126
  indicator 16, 97, 109
  interior 102
  on motorways 109, 113, 117
  traffic 43, 96-7, 199
Limits 57-63
  height 59
  knowledge required 57
  length 63
  weight 58, 247
  width 62
Long vehicles, care when driving 63

MGW (Maximum Gross Weight) 58
MSM/PSL routine 111, 112, 120, 197, 202, 207, 208, 210, 212

253

INDEX

MTVs (Minimum Test Vehicles) 240-1
Maintenance 46-52, 54, 64, 77, 108-10, 122-3 (*see also* Defects)
   disposal of oil etc. 64
Making progress 112, 200
Maximum Gross Weight (MGW) 58
Mechanical failure 11
Medical examination 8
Medical reports 5, 8
Medical standards 9
Medicines 9, 75, 107
'Metro' systems 43-4
Midibuses 33
Mini-roundabouts 213
Minibuses 31-2
Minimum Test Vehicles (MTVs) 240-1
Mirrors 62, 92, 96, 108, 113, 194-7
   nearside 16, 92, 196
Mobile phones 77, 98, 110, 119, 134, 135, 158
Mobile project buses 40
Mobility difficulties, passengers with 20-2
Momentum 28
Motorcyclists 16, 93, 94, 115, 128, 203, 204
   accidents 133, 139
Motorway driving, preparing for 106, 108-10
Motorways, driving on 106-121, 168
   accidents 117, 134
   breakdowns 118-19
   contraflows 118
   crosswinds 115
   end of 121
   fog 117
   fog lights, high intensity rear 109, 113, 117
   frost 116
   hazard warning lights 113, 119
   ice 116
   joining 96, 111
   lane discipline 112
   leaving 120
   overtaking 112, 115
   queues 120, 121
   rain 116
   roadworks 118
   separation distance 113, 116
   signals and signs 114, 117, 118, 120, 121
   slip roads 111, 112, 120
   speed, reducing when leaving 121
   telephones, emergency 119, 134
   weather conditions 113, 115
Muddy conditions 124

Night driving (*see* Driving, at night)
Night vision 101, 102-3

Observation, effective 89, 92-4
Observation at junctions 93
Observation on motorways 111
Oil 47-8, 64, 110, 122
Operator's Licence, PSV 73
Operator's responsibilities 76
Overhangs 35, 63
Overhead clearances 59, 60
Overtaking 16, 214-15
   on motorways 112, 115
   at night 105
   PCVs 91

PAS (Power-assisted steering) 55
PCV driving test (*see* Test, PCV driving)
PCVs, types of 31-45
   articulated 36, 63, 239
   double-deck coaches 38
   double-deck service buses 35
   historic 41
   midibuses 33
   minibuses 31-2
   mobile project buses 40
   playbuses 40
   single-deck coaches 37
   single-deck service buses 34

INDEX

tri-axle 39
Parked vehicles at night 102-3
Parking 82, 84, 126
    brake 52, 53
    at night 102-3
Passenger safety 11, 18, 42, 130, 167
    (*see also* Breakdowns, passenger safety)
Passengers
    blind 17, 20, 21
    caring for 17-22, 30, 89, 134, 168
    children as 18-19, 21, 32, 34, 169
    deaf 17, 20, 21
    disabled 20-2, 30
    elderly 22
    with learning disabilities 22
    with mobility difficulties 20-2, 30
    partially sighted 17, 20, 21
    regulations regarding 85
    sitting before moving off 17, 21
    during test 144
Patience 12, 22, 91
Pedestrian crossings 103, 218-20
Pedestrians 12, 16 92, 93, 94, 103, 132, 203, 204, 216
Pelican crossings 219
'Permit' scheme, vehicles operating under 71
Phones, mobile 77, 98, 110, 119, 134, 135, 158
Physically disabled passengers 20-2, 30
Planning ahead 10, 89, 100
Playbuses 40
Police, notifying 61, 119, 133, 134, 135
Positioning 206-7
Power-assisted steering (PAS) 55
Professional service 19
Professional standards 10, 17, 19
Provisional licence 5, 6, 151, 157, 171
Puffin crossings 220
Pushchairs 21

Radios 101, 110
Rail bridges, collision with 60-1
Rain 116, 123
Ramps 22
Rapid transit systems 43-4
Recovery agencies 130
Red Routes 65, 81-2, 248
    parking and loading in 82
Reflective studs and markings 127
Regular services 71
Regulations, driving 78-83
Regulations, other 73-87
Regulations, vehicle 76-8
Responsibilities, driver's 76, 78
Responsibility to other road users 11
Rest periods 70-1, 74, 100, 101, 107
Restrictions (*see* Limits)
Retaliation 15
'Retarders' 53, 64, 124, 159, 248
Retesting once disqualified 155, 229-30
Reversing, audible warning systems 67
Reversing, exercise for test 176-9, 243
Risk, reducing 130
Road
    camber 59
    conditions 26, 162
    maintenance work 103, 118
    procedure 163-6
    surfaces 26, 212
Road-friendly suspension 66, 86, 246
Roads, congested 16
Roundabouts 210-13
    going ahead 211
    turning full circle 212
    turning left 211
    turning right 212
Route planning 64, 84

Safe distances 95, 100, 105, 113, 116, 118, 123, 126, 202
Safe working practices 86-7, 168-9

INDEX

Safety 42, 86-7
 braking systems 53
 checks 122, 131, 159-60, 163-6, 175, 192
 of passengers 11, 18, 42, 130, 167 (*see also* Breakdowns, passenger safety)
School buses 18-19, 34, 169
Seat belts 32, 34, 35, 169
Secondary brake 52
Security alarm systems 67, 87
Semi-automatic gearboxes 186, 248
Separation distance 95, 100, 105, 113, 116, 117, 118, 123, 126, 202
Service brake 52, 53
Shopping areas, driving in 16, 204
Signalling 16, 97, 198
 flashing headlights 15, 16, 97
Signals and signs 60, 114, 117, 118, 120, 121, 163, 198-9, 225, 244
Single-deck coaches 37
Single-deck service buses 34
Skidding 26, 27
Snow 113, 124-5
Speed 14, 29, 65, 118, 121, 201
Speed limiters 14-15, 78, 79, 112, 159
Speed limits 79
Speeding offences 79
Spray suppression equipment 109, 116, 123
Standards, professional 10, 17, 19
Steering 25
 power-assisted 55
 sudden movement of 25, 26, 27, 118, 123
 use of, for test 190
Stopping at a safe place 222
'Supertrams' 43-4
Suspension
 air 38, 66, 86, 245
  levelling systems 30
 'kneeling' 22, 34, 246
 road-friendly 66, 86, 245

Tachographs 68-9, 71, 72, 159, 192, 248
 charts 68-9
 recording information 69
 symbols 69
Tailgating 13, 95
Technical support 42
Telephones 77, 98, 110, 119, 134, 158
 motorway 119, 134
Television equipment 37, 38, 110
Test, PCV driving
 accelerator skills 183
 address changes 154
 anticipation 203
 application form DLV26: 152, 153
 applying 151-4
 appointments 153, 154
 assessment 147
 attending 171
 awareness 203
 booking 153, 233
 brake control skills 187-9
 braking exercise 180-1
 cancellation 150, 154, 232
 centres 235
 children, legislation re carrying 169
 clutch control skills 183
 controls, knowledge of 159-60, 182-92
 coverage 146, 148
 disability 151
 examiners 146-7, 148, 150
 extended 155
 failing 147, 227
 faults 147
 gear changing 184-6, 191
 handbrake 188, 189
 hazards (*see* Hazards)
 knowledge required 156-7, 159-60, 167-8
 language difficulties 148, 151
 legal requirements 157-8, 173, 226
 licences required 36, 151, 171
 making progress 200
 mirrors 194-7

256

INDEX

motorway driving, knowledge of 168
moving off 193
pass certificate (D10V) 226
passengers during 144
passing 226
photographic identification 171-2
postponing 150
practising, causing a nuisance while 145
preparing for 143-51
ready for 150
report (DLV25A) 226, 227
report form (DLV25) 147
retesting once disqualified 229-30
reversing exercise 176-9, 243
right of appeal 227
road conditions 162
road procedure 163-6
road user behaviour 160
rules, understanding 225
safe working practices 168-9
safety checks 159-60, 175, 192
separation distance 202
signals and signs 198-9
special circumstances 151
speed 201
steering 190
stopping at a safe place 222
syllabus 156-69
Theory 6, 148, 151
topics covered 144, 146
traffic lights 199
traffic rules and regulations 163
traffic signs 163, 225
trailer exercise 177, 223-4
trainer booking 153
training coverage 144
training organisations 143

vehicle
    changes 154
    characteristics 161
    control 163-6
    preparation 172, 192
    requirements 157
vehicles, minimum (MTVs) 240-1
weather conditions 162
Theft 87
Theory Test 6, 148, 151
Timetables 17, 35
Tiredness 74, 100, 101, 107
Toucan crossings 221
Towing trailers (*see* Trailers, towing)
Traffic
    calming measures 62, 65
    lights 43, 96-77, 199
    rules and regulations 163
    signs 60, 114, 117, 118, 120, 121, 163, 198-9, 225, 244
Traffic Area Offices 233-4
Traffic Commissioners 46, 233-4
Trailers, towing 45
    exercise for test 177, 223-4
    licence for 6
Training coverage 144
Training organisations 143
Tramways 43-4
Transmission system 49
Trees, overhanging 59
Tri-axle buses and coaches 39
Tunnels 99
Turbochargers 47, 122, 248
Turbulence, effects of 16
Turning left or right 30, 208-9, 217
Two-second rule 95, 113 (*see also* Separation distance)
Tyre, changing 51-2
Tyre, types of 50-1, 65
Tyre checks 39, 50-2, 108, 122, 167
Tyre failures 27, 39, 66, 93, 108, 131

257

INDEX

Vehicle
  characteristics 161
  control 25, 29, 89, 163-6
  defects 11, 19, 76, 77, 86, 130
  lighting (*see* Lights)
  limits (*see* Limits)
  maintenance 46-52, 54, 64, 77, 108-10, 122-3 (*see also* Defects)
  regulations 76-8
  sympathy 30
  types and licence requirements 241
VOSA checks 46, 77
Video equipment 36, 37, 38, 110
Visibility 113
Vision, zones of 94
Vision at night 101, 102-3
Visually impaired passengers 17, 20, 21

Weather conditions 113, 115, 122-8, 162 (*see also* Fog, Rain, Snow, Winds, high)
  bad 122
  icy 113, 116, 123
  vehicle checks 122-3
Weight limits 58, 247
Wheel checks 131, 192
Wheelchair users 30, 33 (*see also* Physically disabled passengers)
Width limits 62
Winds, high 115, 128
Windscreen 109, 117, 123

Zebra crossings 219

NOTES

Titles for professional LGV and PCV drivers

Prepare for the professional tests and a new career with the official guidance ...

Driving Goods Vehicles – the Official DSA Syllabus

A companion volume to *Driving Buses and Coaches – the Official DSA Syllabus,* covering every aspect of driving a large goods vehicle (LGV). It contains definitive information on the LGV practical test: everything from applying for the test and the latest legislation, through to handling techniques, driving attitude, hazard labels and, of course, the official test syllabus.

ISBN 0 11 552485 1   £14.99

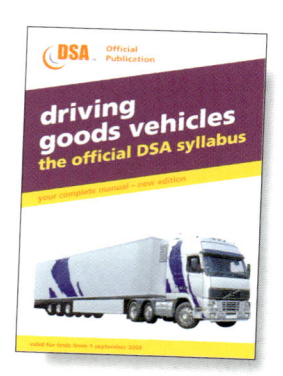

Official Theory Test for Drivers of Large Vehicles

This is the only official theory test book for drivers of large vehicles, both PCVs and LGVs. It contains all the questions you are likely to be asked, plus the answers, along with thorough explanations. Information about test booking and preparation is included, as well as information on where and when the test can be taken.

ISBN 0 11 552451 7   £14.99

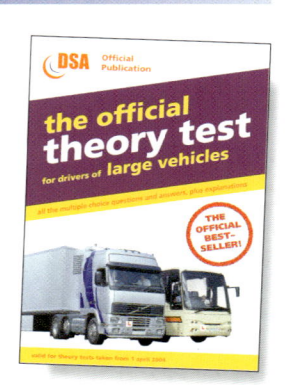